JORULLO

Dr Gadow in Mexico on the journey home.

JORULLO

THE HISTORY OF THE VOLCANO OF
JORULLO AND THE RECLAMATION
OF THE DEVASTATED DISTRICT BY
ANIMALS AND PLANTS

By

HANS GADOW, M.A., Ph.D., F.R.S.
LATE STRICKLAND CURATOR AND READER
IN VERTEBRATE MORPHOLOGY IN THE
UNIVERSITY OF CAMBRIDGE

CAMBRIDGE
AT THE UNIVERSITY PRESS
1930

CAMBRIDGE UNIVERSITY PRESS
Cambridge, New York, Melbourne, Madrid, Cape Town,
Singapore, São Paulo, Delhi, Mexico City

Cambridge University Press
The Edinburgh Building, Cambridge CB2 8RU, UK

Published in the United States of America by Cambridge University Press, New York

www.cambridge.org
Information on this title: www.cambridge.org/9781107670426

© Cambridge University Press 1930

This publication is in copyright. Subject to statutory exception
and to the provisions of relevant collective licensing agreements,
no reproduction of any part may take place without the written
permission of Cambridge University Press.

First published 1930
First paperback edition 2013

A catalogue record for this publication is available from the British Library

ISBN 978-1-107-67042-6 Paperback

Cambridge University Press has no responsibility for the persistence or
accuracy of URLs for external or third-party internet websites referred to in
this publication, and does not guarantee that any content on such websites is,
or will remain, accurate or appropriate.

CONTENTS

	PAGE
Editorial Note. By Philip Lake	vii
Prefatory Note. By A. C. Seward	ix
Author's Preface	xi
Jorullo and its Surroundings. (A preliminary sketch.) By Philip Lake	xv
Chap. I. The Story of the whole Drama	1
II. The Process and Progress of Reclamation of the Devastated District. (a) Plants	22
III. The Process and Progress of Reclamation of the Devastated District (cont.). (b) Animals	49
IV. The Amphibia and Reptilia of Michoacan	65
Appendix. Extracts from the Literature concerning Jorullo	76

ILLUSTRATIONS

Dr Gadow in Mexico on the journey home *Frontispiece*

Fig. 1. Jorullo from La Playa, with the Malpais in front. Sketch by Mrs Gadow . *to face p.* 1

Fig. 2. Jorullo from the Mata de Plátano. Sketch by Mrs Gadow . *to face p.* 1

Map of Jorullo
 available for download from www.cambridge.org/9781107670426

EDITORIAL NOTE

THE MS of this work was left by Dr Gadow in a condition almost ready for press. A few references were missing, a few small points remained to be cleared up, and Dr Gadow had made notes for a final revision. With the help of these it has been possible to complete the MS nearly in the form in which he would himself have put it, excepting only that he had also intended to insert the genus and species of the plants to which he refers by their Mexican names. Where his emendations could not be incorporated verbally a note has been added giving their substance. To distinguish these and other explanatory notes from those of Dr Gadow himself such notes have been initialled.

The references have been completed and obvious slips have been corrected. In order to render Dr Gadow's story of the eruption more easily intelligible to those unacquainted with the district a short preliminary description has been prefixed, and for the same purpose occasional small verbal changes have been made in the text.

The illustrations of the volcano and its surroundings are from sketches by Mrs Gadow. In the map which has been added, the geological lines are from the map by Ordoñez, the contour-lines from that by Villafaña.

Dr Gadow contemplated the preparation of a second volume on Mexico, in addition to that which he had already published. It seems likely that after he had finished this MS he decided to keep it back for inclusion in the larger work.

PHILIP LAKE

PREFATORY NOTE

THOUGH unfamiliar with the district described by Dr Gadow, my long friendship with a former colleague and my interest in the facts here recorded make it a pleasure to accept the invitation of Mrs Gadow to contribute a prefatory note. Dr and Mrs Gadow, in the course of their journeys in several little-known parts of the world, have added greatly to our knowledge of natural history, especially of certain districts of Mexico. Jorullo affords a particularly instructive example of what may be called a large-scale experiment in Nature's laboratory—the destruction of vegetation over a wide range of country and the reintroduction of animals and plants by natural agencies. It tells much the same story as the volcanic island of Krakatao, which is a classical example of re-colonisation following devastation. In a recently published book (*The Problem of Krakatao as seen by a Botanist*) summarising the results of several expeditions to the island since the catastrophe of 1883, C. A. Backer puts forward a view which is opposed to that usually held. Prof. Treub, a former Director of the Buitenzorg Botanic Garden in Java, states that the terrific explosions and showers of ash of August 1883 had destroyed all animal and plant life. Botanists generally accepted his statement and regarded the present flora and fauna of the remnant of the devastated island of Krakatao as new introductions from neighbouring lands. Mr Backer suggests that the destruction was not complete, though it may be assumed that many of the plants which have re-clothed the ash-covered remains of the island were derived from external sources.

Jorullo differs from Krakatao in several important respects. It is not an island; the eruptions continued for some years; and great outflows of lava covered an area of about five square miles. There can be no doubt that throughout the lava-covered region all life was completely destroyed; but the destructive effect of showers of ash extended far beyond the reach of the lava. The total area in which life was completely or almost completely destroyed greatly exceeded that at Krakatao. But at Jorullo there was no barrier of sea to hinder the re-entrance of plants and animals.

A. C. SEWARD

AUTHOR'S PREFACE

WE went to see Jorullo and our visit extended to a whole month, so we did him thoroughly. Our special problem was to study the manner in which the devastated district had been reclaimed by the surrounding fauna, and perhaps to find the rate at which it had been, or was still, proceeding, because this is a factor about which very little is known. It may seem unnecessary to bring such a large apparatus to bear upon a question of zoological distribution, but the fauna is absolutely dependent upon the flora, and this depends upon the kind and condition of the ground. Here was a district, of more than fifty square miles, so fertile until the shock of September, 1759, that it was known by the Tarascan name Jorullo, the Paradise. In the midst of it a series of volcanoes arose and covered a square of several miles with lava, whilst for many miles around the country was smothered with volcanic ashes and sands. To tackle the whole problem of reclamation it was necessary to study the very varied features of the terrain, as to subsoil, water-supply and drainage, the probable roads of ingress taken by the new colonists, and, to arrive at an estimate of the devastation, to get a *tabula rasa*, a clean slate. I was familiar with Humboldt's account, and we had with us the pamphlets by Ordoñez and Villafaña published on the occasion of the Tenth International Geological Congress, a party of which paid a hurried visit to the Jorullo about 21 months before we appeared upon the scene. The accounts hitherto published do by no means agree—the history of the whole catastrophe is singularly garbled, and to us at least it seemed remarkable that even the Mexican authorities have scarcely attempted to sift critically and correlate the various printed accounts. No use has been made of the diary written by an educated man, an exposition of what happened during the first six weeks, although this diary has been reprinted by Villafaña in his most interesting pamphlet. Since it is written in Spanish, a language not very familiar to most readers, I have translated it.

We had plenty of time in camp, during our halcyon days at Mata de Plátano, and whilst roaming over the district, to ponder over the various points of discord, to verify this and to question that, until the whole drama seemed to re-enact itself. It would have been futile to trace the fauna on to Jorullo without knowing something of the fauna of the surrounding neighbourhood, and soon this task resolved itself into a chase which led to the very delta of the Balsas on the shore of the Pacific, and thence back, crossing the western half of the same volcano-studded depression of which the Jorullo district forms the eastern portion, and finishing up, rather abruptly with fever, on the much older Volcan de Tancitaro, which rises to beyond 13,000 feet.

Zoologically the southern half of the State of Michoacan was *terra incognita*—there has been some butterfly hunter and a collector of orchids. For the special purpose we had in view I have restricted myself almost entirely to Reptilia and Amphibia on account of their superior importance for questions of the mode and rate of distribution. Mammals are too few and to be got only by laborious trapping, and they roam about too easily. Birds are hopeless, fishes too few—only five species have ever been recorded by ichthyologists from the whole of the Balsas basin, and from the Jorullo there was none. Reptilia and Amphibia alone are bound to their place; they do not migrate.

HANS GADOW

JORULLO AND ITS SURROUNDINGS
(A PRELIMINARY SKETCH)

UNTIL A.D. 1759 there was no hill where the Volcan de Jorullo now stands, nor were there any records of volcanic activity in the neighbourhood. The inhabitants did not even know what lava was. There were, indeed, volcanic hills close by, but these had long been extinct and had no meaning in their eyes. The eruptions, which began in 1759 and continued intermittently for several years, completely buried a tract of more than five square miles which had previously been famous for its fertility. The destructive effects extended over a very much wider area, and even at Queratero, 125 miles distant, ashes fell in abundance.

Historically the volcano is interesting because Humboldt, who visited it in 1803, described it as an elevation crater, formed in accordance with the theory of Von Buch. That theory has long been abandoned and the real interest of the eruption lies in the fact that full reports of the birth of the volcano, written by an eyewitness, were sent to the Viceroy at the time and were preserved in the Archives. These reports summarise the premonitory signs which had been observed for several months, and give a diary of events from September 29th, when the eruption began, to November 13th, on which date the second report closes. The reports were unknown to Humboldt and were first published in 1854 by Orozco y Berra. They have since been twice reprinted, but Dr Gadow seems to be the first who has seriously discussed their significance.

Jorullo stands in an amphitheatre amongst the foothills of the great southern slope of the Mexican plateau. The amphitheatre is roughly rectangular in shape, with its longer sides running from east to west and its shorter sides from north to south. It was closed on north, east, and south, but opened westward into the plain of La Playa and Agua Blanca, the opening being divided by the Cerro del Veladero. The amphitheatre is now filled by the products of the eruptions and its original floor is nowhere visible, but a single hill which stood in its midst still raises its summit through the lava flows. This is the Cerro Partido, and according to tradition the old Hacienda of Jorullo was situated in its vicinity.

The seat of the eruptions lay along the eastern side of the amphitheatre, where four distinct cones may now be seen, viz. Jorullo itself, the Volcancito del Norte to the north of it, and the Volcancitos de Enmedio and del Sur to the south (Fig. 1). A line joining the summits of these cones is nearly straight and runs approximately from east-north-east to west-south-west. It indicates, no doubt, the course of the fissure from which the eruptions took place.

The three Volcancitos and the southern half of Jorullo are formed, superficially, of volcanic ash and agglomerate; but the northern slope of Jorullo is covered by a sheet of lava proceeding from the present crater. Towards the east the cones abut against the hills which shut in the amphitheatre; but towards the west they slope down to an extensive field of lava which fills most of the amphitheatre and which is now known as the Malpais.

JORULLO AND ITS SURROUNDINGS   xvii

According to Ordoñez the lava flows I and II, shown upon the map, were first extruded. An explosive phase followed and the four cones were built up. At a later date the smaller flows III and IV came from the vent of Jorullo. Dr Gadow believes that the first phase of the eruption was entirely explosive and that even the oldest flow did not take place till later. He was led to this view chiefly by the reports of Sáyago, to which reference has already been made. Sáyago, who was an eye-witness, describes showers of ashes, sometimes incandescent, and flows of sand or mud, but in his first report makes no mention of lava. The Viceroy evidently noted this omission and asked whether there were not flows of lava. In his second report Sáyago tries to answer the question. He says that he does not know what lava is like and has no one who can tell him, but he states definitely that nothing flows except the ashy material which he describes. It seems quite clear that up to November 13th there had been no great outburst of lava. Dr Gadow suggests that the flows began in 1764, when, according to common report, the eruptive activity reached its maximum.

In order to avoid danger of confusion it may be useful here to point out the differences between Ordoñez, Villafaña and Gadow with regard to the number and delimitation of the lava flows. Ordoñez distinguishes four, which are shown upon the map, numbered in order of succession. Villafaña recognises five, his IV being the portion of Ordoñez' III which is nearest to Jorullo, and his V the equivalent of Ordoñez' IV. Gadow gives no map of the flows, but from his description it is clear that his fifth flow, which he thinks came from the Volcancito del Norte, is the semi-detached northernmost portion of Ordoñez' II. His sixth flow forms the Calle de las Ruinas, a conspicuous trough-like channel which starts northwards from the Jorullo crater and bends towards the east, widening and shallowing as it goes. The Mexican geologists consider this as merely the last phase of the flow that covers the northern slope. A thin stream of lava continued to run beneath the crust after the rest of the flow had solidified, and the roof afterwards fell in.

P. L.

Fig. 1. View of Jorullo from La Playa, with the Malpais in front, the Volcancito del Norte to the left of the main cone and the Volcancito de Enmedio on the right. Sketch by Mrs Gadow.

Fig. 2. View of Jorullo from the Mata de Plátano. Sketch by Mrs Gadow.

Chapter I

THE STORY OF THE WHOLE DRAMA

THE story of the whole drama may be rendered as follows.

Towards the end of the month of June, in the year A.D. 1759, the people living at Jorullo were alarmed by subterraneous noises and knocks which off and on repeated themselves until September 17th, when the noises sounded like the discharge of cannon, and the earth trembled to such an extent that the chapel of the main hacienda was badly cracked. These disturbances kept on with scarcely any intermission until the 27th. This and the following day passed quietly, but when some labourers returned from gathering guava fruit, ripe at this time, they caused astonishment because their sombreros were covered with dust or ashes. Unless this little episode is a fiction, some crack in the ground must have opened and have exhaled this dust quietly without attracting attention.

On September 29th at 3 a.m. were felt several sharp tremors, and from the bottom of a ravine about half a mile to the south-east from the hacienda broke out dense and dark steam, soon followed by roaring flames, and there arose into the sky a thick and dark cloud. The discharged superheated steam condensed, and fell down in the shape of rain, which, mixing and carrying with it the likewise expelled sandy cinders and ashes, soon covered the neighbourhood with mud. This must have happened soon after the first outburst, as the people on leaving the chapel, into which they had been summoned to hear Mass, found the buildings already loaded with mud. By the end of the day the farm buildings were laid low and the whole farm was entirely spoiled by the falling masses of mud, namely the rain mixed with the ashes. For two days the newborn volcano thundered and threw up sand and fire without one minute's interruption.

On October 1st two new things happened. First, from the foot of a hill, a little to the south of the volcano, burst forth a current of muddy water, voluminous enough to prevent one crossing it. Second, a mass of sand rose to the outlet of the volcano, which at that time was little more than a cleft, and flowed into the bed of the Cuitinga brook; but this sand was dry and so hot that it set on fire everything in its path; having followed and filled up the brook for about half a mile, the water underneath exploded at several places, throwing torn sods high into the air.

During the night of October 2nd happened a serious earthquake, followed by fiercer outbursts of ashes which by the 4th smothered the western part of the district to a distance of 5 miles and even beyond, so that by October 6th the whole population of La Guacana (about 5 miles to the west of the volcano) had to retire on to the neighbouring hills. Things were aggravated by a new kind of danger. The little stream from the Jorullo to La Guacana became so swollen, and apparently blocked with the masses of sand and ashes which it gathered in its course, that all the lower parts of the valley, the present Playa, the now level stretch to the west of it, and thence extending towards La Guacana, were transformed into a lake. It is worth noting that this inundation was caused not only by the water which fell as rain, but to a much greater extent by many new streams which burst from the neighbouring hills, now swelling suddenly into regular spates and then again suddenly running dry. It has been suggested that it was all rain water, real rain, and that tropical thunderstorms caused these torrents to come from the slopes of the wide amphitheatre of the whole district; but by the beginning of October the rainy season is practically over. It begins with regular downpours in June, reaches its maximum in July or August and in the following month tails off into occasional smart showers. The whole annual rainfall in this part of Michoacan is not at all 'tropical', amounting perhaps to 1 metre. The eruption may have changed all this, but against it is to be noted that the careful diary of the gentleman who witnessed the phenomenon does not contain a single reference to thunderstorms except on and after November 8th—at that time of year something very exceptional and therefore duly mentioned.

Incessant falls of ashes seem to have continued until October 8th, when a new phase began, the volcano throwing up stones, very brittle, as if overbaked and glassy, and these fell to a distance of a mile from its mouth; the ashes had of course been carried much farther, by the wind—for more than 50 miles.

During the night of October 9th were felt strange shocks, accompanied by great noise, and during the following two days there fell again great masses of ashes and rain from the globular cloud, but in addition to these the volcano threw out many incandescent bombs, some of the size of the trunk of an ox.

On October 12th, at 1 p.m., 600 yards from the main crater, a new mouth opened which extended over the whole gorge to the west; out of this came a thick cloud of steam together with such a volume of hot water that it flowed for two hours like a spate, whereupon the gap closed and the water ceased.

On October 15th things were quiet enough to allow the Administrator with his labourers to go to the chapel and rescue the holy images and save the bells from the tower. The same kind of rescue work was undertaken some 10 days later by the priest at La Guacana.

From October 14th a gradual improvement began; the volcano no longer threw out steam, but only dry ashes together with mighty flames of fire which roared like billows, the noise changing to thunder during the occasional discharges of stones. But there was no longer any rain of 'mud' and the springs of water had run dry, and on November 1st the sun came out, soon to be drowned again by renewed showers of ashes. From November 9th to November 12th the darkness was worse than ever, accompanied by furious earthquakes and several hurricanes with thunder, lightning and downpour of rain all over the neighbourhood.

November 13th was quiet. The whole of the more fertile slopes and plains was thickly covered, and partly levelled, with the fallen ashes and sand, so that all the old brooks were running on the top of the new level and quite clear. The volcano had piled itself up to a height of more than 300 varas (about 250 metres or 820 feet) with a circular crater. The careful recorder, Administrator Sáyago, had even taken the trouble of making a sketch in black and red of the volcano as it appeared on

October 8th. Orozco, who saw this most interesting document in 1854, dismissed it with the withering remark that it was executed with little care and scarcely served to convey an approximate idea of the object! This is the criticism of a person who had never visited the Jorullo, who therefore could compare the sketch only with those published by Humboldt or Schleiden. Of course the sketch of October 8th, 1759, cannot bear any but a slight resemblance to the later features, because not only has the volcano become much higher but there was not yet, in 1759, that enormous outflow of lava, which formed the Malpais and has thereby changed completely the aspect of the whole district.

After the Administrator Sáyago had dispatched his report of November 13th to the Governor of Michoacan, we hear no more of him. Naturally not; since he had done his duty and the ruined farms had to be abandoned, he left the district. There is a letter by the Viceroy Amarillas to the Minister of the Indian department in Madrid, dated Cuernavaca, November 21st, 1759, in which the Viceroy reports that he has given orders to the authorities to send him further accounts of what may happen at Jorullo and also to resettle the natives at La Guacana. But henceforth there is a complete absence of news, except oral traditions that violent eruptions continued until February, 1760, again during the next 4 years, and others apparently less great for 11 years more, bringing the end of eruptive activity down to the year 1775. There is, further, the allusion by Clavigero to an account rendered to or by the new Governor Bustamante in 1766, according to which three high mountains had formed themselves at Jorullo with a total circumference of 6 miles, a very fair statement of the actual conditions, the straight distance from the Volcancito del Norte to the V. del Sur being about 2 miles. We can therefore safely conclude that the main features of the Jorullo district as they are now, namely several volcanoes and the Malpais to its present fullest extent, had been completed by the year 1766.

The most interesting of these features is of course the great Malpais, the outbreak of lava, which covers several square miles to a depth of perhaps more than 100 metres, at least in the centre. All this must have happened between, say, the beginning of the year 1760 and 1766, and probably began in 1764, the period of alleged greatest eruptive activity.

This spreading of molten lava over whole square miles of land must of course have been a grand spectacle and I have not the slightest doubt that it is to these scenes that those stories refer which Colonel Riaño and Fischer were told on the occasion of their visit in the year 1789; and the same stories were told to Humboldt, 43 years after the event, when the number of surviving witnesses had probably as much decreased as tradition had adorned the facts. Be it noted, however, that the spectators observed what they saw from the heights of Aguesarca. Humboldt himself mentions this place in his account, but does not tell us that the distance to the Jorullo is about 6 miles as the crow flies. The tradition is good enough. A large stretch of land was seen to be in a frightful outburst of fire, and in the midst of the flames there appeared a large shapeless lump, like unto a black castle. Or, to give Pieschel's version (p. 96), from half a German square mile arose flames which threw up stones and thick smoke, and the molten soil rose like an agitated sea. The Indians may even have observed the formation of huge bladders and blisters which afterwards burst.

All these may be taken as facts; only they did not happen in the night of September 29th, 1759, but long after the volcano had built itself up to a height of some 250 metres; nor did they happen on one day, but they spread perhaps over years. Whatever has been added to the story—the perpendicular lifting up of the ground, its swelling in the middle and at several other places, then bursting and thus producing the several volcanoes —is pure invention.

Humboldt was naturally enchanted with his notion of having found, in Jorullo, an example of the elevation theory of his great friend L. von Buch, but it is nevertheless very remarkable that he does not say a word in his description of the Malpais about its being composed of lava. He makes much of the abruptly rising edge of the Malpais where it faces La Playa, and describes it as composed of banks of black and brown clay, covered on the top with but little volcanic ash. Burkart, who may have examined the edge at some other spot, found the wall to consist of light grey, not at all dense, basaltic rock and divided into several banks by undulating, almost horizontal cracks. Schleiden found it made up of blocks (of basalt) mixed with layers of 'schalen'.

Although not a geologist, I venture to give what I have observed myself. All the previous descriptions are right. It depends upon the place where the edge of the Malpais is examined, and its sharply upstanding or abruptly ending margin is several miles long. At some parts, for instance in the north, between Cerrito de la Cruz and the Cerro do Paso Hondo, it is now almost level with the rest of the ground, because washed-down sand and fallen blocks have levelled the difference; but opposite the last-named the edge rises to a height of perhaps 50 feet, quite impossible to climb and bare of vegetation because some of the upper ledges, especially the top ledge, are overhanging. There are many such ledges, maybe a foot thick, of hard blistered blackish lava, separated by layers of red or brown earthy matter, here and there with streaks of paler lapilli or of black, finer ashes or sand. Rarely are these strata horizontal, but mostly undulating or faulted. The soft layers are crumbling away, leaving the harder ledges between them. The deeper down they go, the more sand, or ashes and clay-like soil; higher up, the stony, basaltic matter becomes preponderant.

These conditions are still more striking a few hundred yards to the south of La Puerta, where the little stream has by this time made itself a deep gorge, having cleared it of all the once overlying sand so that its bed is now formed by the old crystalline rocks which underlie the whole district. Here and there the cliff rises over 50 feet and its face presents regular caves. Near the foot is non-stratified soil, which may be the original surface soil of the plain before there was any volcano. Then follow layers, or, as they present themselves in concave vertical sections, masses or nests, of other stuff, sandy, ash-like, mixed with soil, with inextricable contours; and still higher up follow the same alternating layers of earthy (i.e. disintegrated) matter, ashes, stones and strata of lava as I have described above. It may be an instance of unprofessional rashness, but the masses of about the middle third gave me the impression as if I were gazing at the 'mud' which is reported to have rained for weeks or which was carried by the spates which burst from the newly born volcano. According to the slope of the whole terrain they would have flowed in this direction (just like lava flow No. 1), blocking and filling up the whole gorge and thereby causing the great inundation.

The non-historic building up of the volcanoes and the Malpais, after November, 1759, seems to have proceeded as follows. We know there was a deep ravine at the bottom of which was the bed of the Cuitinga stream, along the foot of the basaltic hills ('basalto de nephelina' according to Ordoñez). This ravine we may well assume extended from the present Volcancito del Norte in an almost straight line from north-east to south-west to the southern volcancito, whence—according to the old diorite hills of the Peñablanca and the Cerros de las Pilas—it turned about at a right angle to the west, the stream passing between the Veladero and Agua Blanca, below which it joined the stream coming from the Playa. We know further that the future main volcano broke out at the bottom of the Cuitinga ravine. This was repeated, about 1400 metres to the north-east, by the future Volcancito del Norte, and to the south-west by several other volcancitos, the V. del Sur arising at a distance of 1600 metres from the main volcano. According to Ordoñez the Sur was the latest, because it has disturbed, or overlaid, the southern slope of the Enmedio.

It is important to note that there were most probably more than two southern cones. There is an almost comical uncertainty about the total number of cones. Sáyago, the eyewitness of the first eruption, describes only one, although the following passage may bear various interpretations: 'having belched out such a mass of red-hot stones, so that around its mouth was formed a circular wall, which is already higher than 300 varas and surpasses the others which stand on the sides of the ravine which latter it has filled up and disfigured'.[1] Does this mean the other volcancitos which stand at either end of the ravine, or does he refer to the neighbouring foothills and to the Cerro Partido which at that time rose about 200 metres above the original plain?

According to Landivar the big volcano had four companions, five in all.

Clavigero (*fide* Bustamante) said that three high mountains were formed; he may have omitted the smallest.

Humboldt counted and sketched six. Felix and Lenk likewise counted five 'Nebenkrater', six in all.

Schleiden drew four; and Ordoñez insists upon four only, viz. Jorullo itself, the Volcancito del Norte to the north of it, and the Volcancitos de Enmedio and del Sur to the south.

To ourselves it was obvious that there are five, the fifth adjoining the Enmedio but partly covered by its slope, although a little higher. It is unmistakable when examined from the top of the Enmedio, and to anyone walking on it there appear good traces that it was

---

[1] '...de modo que en el recinto de su boca ha formado un brocal, pretil ó círculo, que ya pasa su altura de trescientos varas, y sobrepuja los demás que están á los lados de la cañada, la que totalmente ha llenado y desfigurado.'

breached exactly like the others. But its contours are more rounded off by the masses of sand on the top and on its slopes. In fact it has been left more smothered than the Enmedio, just as this is more disfigured than the Sur. Ordoñez' chief objection to its being a volcanic cone is his discovery of undisturbed, or non-dislocated, old nepheline basalt near its foot. One of such patches is indicated in Villafaña's map about 300 metres to the east of the Enmedio, and another towards the northwest. The distance of this questionable volcancito is about 800 metres from the main volcano, space enough for still another volcancito to be hidden completely beneath the south-south-west slope of the big cone. Humboldt has put it in its right place in his little *croquis*, but his sixth cone is a small ash-covered hill at the foot of the north-east slope of the V. del Norte.

However, no matter whether there are, or were, four or five or even six cones, they all lie in an almost straight line, all of them were breached alike on their south-west side, pouring out masses of lava, so that taken together, or considered from a broad point of view, they formed for all practical purposes one eruptive fissure about 4000 metres long. It is immaterial whether this fissure ever existed in its total extent at the same time. After it had opened in the north, El Norte piled itself up, was breached, poured out its share of lava, etc., and then closed up. It is certain that such was the series of events at the main volcano. After it had piled itself up to a height of 300 yards, it poured out a big flow of lava, then it piled itself up to its present height and then it was again breached in the north by the latest lava flow of all. Most of the lava which has formed the Malpais seems to have come from the cleft, or series of clefts, to the south-west of the main volcano.

Humboldt, in support of his hypothesis of the bladder-like mode of elevation, has repeatedly laid stress upon the convexity of the Malpais; cf. also his profile (*Atlas pittoresque*, figs. 19–21). This is quite erroneous. Instead of being convex, with an apex somewhere in the centre, or at the foot of the volcano, the Malpais consists of a number of terraces, steadily descending from the Volcancitos Sur and Enmedio towards the north-western corner of the Malpais. It is easy to construct profile sections from Villafaña's careful hypsometric survey. Such a section from the volcancitos, where the old basalt crops out, to La Puerta, shows a very steady fall from an absolute level of about 980 metres to 770 metres; the highest point of the lava field itself, near the western bases of the cones, is about 950 metres, and this cannot be much above the original bottom. But the surface, rather level, in the corner formed by the Cerro Partido and the foot of the main volcano, is 930 metres, and probably more than 100 metres above the old bottom. The first flow must therefore have come from the cleft upon which now stand the southern cones.

Ordoñez and Villafaña do not quite agree as to the numbering and counting of the lava flows.

The first flow was by far the biggest, and the second is a shorter and later edition of it, and was diverted, or partly stowed up, by the Cerro Partido. This is quite a small hill, but rather classical, because near it the old hacienda is reported to have stood before there was any volcano. Before being buried to the extent of several hundred feet in lava, it must have been quite a feature arising in the midst of the fertile plain. Now it is perhaps only 100 feet high, with a curious saddle-shaped depression, hence its name of the 'parted hill', tradition saying that it was rent asunder during the paroxysm of the birth of the big volcano. Its slopes are very steep and, especially on the southern side, deeply covered with loose yellow ashes. The top is very uneven. This hill consists, according to Ordoñez, of dolerites. A gentleman of the Inguaran Company told me that they had worked it for a while for silver, with some success.

The third and fourth flows, both to the north of the Cerro Partido, appear to be emissions from the main volcano, very steep terraces, at some parts 30 to 50 feet high, marking the edges of the superimposed cakes. The face of these cliffs is extremely rough, full of projecting ledges, or covered with tumbled-down hummocks. A fifth flow[1] belongs to the Volcancito del Norte. The sixth and last[2] has breached the main volcano on its northern rim, thence spreading out in the northern quadrant and to the east. This sixth flow is the latest of all; this alone is quite black, bare without any ashes, and quite barren, a most impressive-looking mass of blocks, tumbled, squeezed and cracked. After it had been vomited out there followed no more of those rains of sand and ashes which still cover the whole Malpais and all the other cones, except where they have been blown or washed away. Even the cone of the big volcano (but not the latest outflow of lava) had once been covered

[1] This is not No. v of Villafaña's map, which comes from Jorullo itself. Neither Villafaña nor Ordoñez recognise a distinct flow from the V. del Norte.—P. L.

[2] Considered by the Mexican geologists as only the final phase of the latest flow.—P. L.

with ashes, otherwise Humboldt could not have spoken of it as 'der weisse Aschenhagel'. Now these pale masses of ashes and sand have been washed off, except what remains in the deep cleft-like valleys between the numerous ridges which radiate from the top on every side of the cone and give it such a peculiar appearance. The last flow still looks so fresh that Hobson felt almost inclined to consider it as of comparatively recent date. However, it was there already at Humboldt's time, just as it stands now; and no serious activity is said to have been noticed after the year 1774.

The first description of the *crater* is by Fischer, who examined it in the year 1789. He ascended by the eastern slope. 'Arriving on the top one passes a kind of flat, full of fissures, a foot or more in width, whence issues smoke and steam. This flat space forms the margin all around the crater, the throat of which is quite involuted and surrounded by vertical and overhanging walls, which are incrusted with yellow and white matter and smoke incessantly. The width of the crater measures 800 feet[1] from S. to N. and 400 ft. from E. to W. There is no lava proper, but half-molten stones baked together by various salts. Towards the west [of the volcano] are several spots still on fire, and towards the end of the devastated district, called the Bad Land, many boiling hot springs are met with.' Ordoñez gives 520 metres and 385 metres as the dimensions of the crater; considerably less than Fischer's superficial estimate.

The depth of the crater was judged by Humboldt as 280 feet, the deepest part reached by him being 140 feet, whilst the deepest spot visible was only as deep again

[1] *Schuh*, i.e. feet, is obviously an error instead of *varas* or yards.

('nur noch einmal so tief'). Felix and Lenk, in 1888, measured a greatest depth of 132 metres below the highest point of the rim, or 110 metres below the average level of the rim. Ordoñez, in 1907, returned the difference between the highest spot, Riaños peak, and the bottom as 149 metres.

Since Felix and Lenk, like Burkart before them, most likely by some error, had found the absolute height of the crater some 80 metres lower than Humboldt, they suggested that the height had diminished, owing to the falling-in of the walls of the crater. Strangely enough Ordoñez objects to this explanation upon the ground that the fallen portions would have filled up and merely elevated the bottom![1] The real explanation is, I think, the following: What Humboldt could not descend into, had become accessible some 80 and 100 years later. Further, allowing for errors, and taking an average top level, the depth has increased steadily within the last 100 years, say from 300 to well over 400 feet. Obviously the bottom has sunk in, and it is still sinking. The flat space over which Fischer walked, full of fissures, has disappeared. Schleiden found the eastern rim, for a length of 30 steps, reduced to a wall only 2 feet wide at the top. Felix could no longer walk this ridge, nor could any of our own party, the wall now breaking up fast from within outwards. But Pieschel makes the very proper remark that the crater having sunk in, must therefore also have widened, and this was indicated not

[1] Ordoñez' objection is based upon the fact that, in spite of this, Felix and Lenk found the difference of height between the floor and the rim of the crater greater than Humboldt did. Humboldt's figure for the highest point is 19 metres less than that now assigned to it (1320 metres).—P. L.

only by the amphitheatre-like terraces but also by the fact that here and there 'some tree or shrub is growing upon the inner side of the crater wall, which must have slid down with the ledge from the outer rim, especially since, but for these trees, the crater is still absolutely bare of vegetation'.

All this we could corroborate. Especially on the southern, south-western, and on part of the western sides the very rough surface looks like a pie which has sunk in, cracked into many more or less parallel fissures, widest at the surface, and not a few of the trees, which had been growing straight, are now inclined inwards, whilst the roots of some are broken and others are actually still spanning the fissures! Thus this whole portion which originally formed the outside (or maybe a remnant of Fischer's flat) now slopes into the crater, or rather towards it, the real crater falling off rather abruptly. All this implies of course that the crater is becoming wider and deeper, and, rather paradoxically, that it also becomes lower, at least at some parts. The process is still going on. Noises, rumblings, have often been reported as proceeding from the volcano. We ourselves heard none during our month's stay in its immediate vicinity, but the bailiff of Mata de Platano, a reasonable man of forty, described them as of something falling, muffled, but without tremors.

The best description of the much-discussed *hornitos* is perhaps the one given by Humboldt in a letter of August, 1857, addressed to Beyrich (*Zeitschr. Geol. Gesellsch.* IX. pp. 297–9), with the reproduction of a sketch made on the spot in the year 1803, and an extract from Humboldt's original note-books. 'The many thousands of small, crater-less, conical elevations (in reality not quite round, but of a somewhat elongated, oven-like shape) which cover the elevated, lifted-up plain rather uniformly, are from 4 to 6 or 9 feet high. Nearly all of them have arisen at the western side of the big volcano. Every one of the many hornitos is made up of decomposed balls of basalt, with concentrically shell-like pieces, often with 24-28 such shells. Most of these balls are somewhat flattened, with a diameter of 15 or 18 inches, with extremes from 1 to 3 feet. The black mass of basalt is perforated by hot vapours, and dissolved into earthy matter; only the core is denser, whilst the shells, when peeled off, show yellow spots of oxydised iron. Curiously enough the soft clay-like mass which joins the balls together is also divided into curved lamellae which are twisted through all the interstices between the balls. It has occurred to me whether the whole might not be one mass which was disturbed and arrested in its formation, instead of its being composed of disintegrated balls of basalt containing some olivine. Some of the hornitos are so much dissolved or contain such large inner cavities, that a mule breaks into them with its fore-feet, an experience which does not apply to the hills of termites. I have found neither clinkers nor fragments of older rocks baked into the basalt of these hornitos, as is the case with the lava of the big volcano. The name *hornos* or *hornitos* is well justified since the vapours escape from the sides instead of from the top. In the year 1780 one could still light a cigar by digging it 2 or 3 inches into such an oven, and some parts of the ground were still so overheated by them that one had to make a detour in walking across the plain. According to the testimony of my Indian companion the place has cooled down during the last 20 years, but I still found a temperature of 93–95° C. within the clefts of the hornitos and that of the air 20 feet away from them measured up to 46·8° instead of only 25° as at La Playa. To call them cones of eruption instead of elevation might easily convey the erroneous notion that there were indications of these hornitos having thrown out clinkers or having poured out even lava like so many cones of eruption.'

It may seem like a storm in a tea-cup that these little hornitos should have caused so much discussion, and without any satisfactory result, and yet I think their mystery can be solved without doubting the veracity of any previous writers or hurting the feelings of any professional geologist. We, my wife and I, had plenty of time during our four weeks' sojourn at the Jorullo to look for their remains, to think and to talk about them.

First. To the west of the foot of the big volcano and south of the Cerro Partido, where the ground is thickly covered with sand and ashes to the depth of many feet, there are many slight rises or hillocks, only a few feet high, sometimes not larger than a cartload of sand dumped down and then smoothed over. In the middle of some crops up a core of rock; the whole thing often sounds hollow and it is made up of rings and rings of different colours, black, yellow and reddish ashes and sand, these layers varying from the thickness of a book-cover to half an inch and more. On a flat view, and where no vegetation exists to disturb their arrangement, these rings appear much broader, sometimes one foot asunder, owing to the slanting section in which they are exposed. They give the impression of wet layers of sand or ashes having been heaped upon a core, layer after layer, and then having been baked together. The ground between these little hillocks, likewise sand or ashes, is not stratified, an indication that it has accumulated since the hillocks were formed, which in fact have partly been buried and at the same time have fallen away in the usual process of ground-levelling. These things, then, I take to be the remains of buried hornitos, which if excavated would probably look very much like the reconstructed objects figured by Schleiden.

Second. About a quarter of an hour's walk to the east of the little village, La Puerta, upon the Malpais itself (still called Malpais because it is either too rough, or too much covered with black ashes, to be fit for agriculture or for the grazing of cattle), are many lumps standing out above the sand-covered ground like glacial boulders. They stand 5 to 8 feet high, not cones but rather box-like, of black basaltic lava, just like the latest flow from the volcano, very rough; sometimes, or in part, like a heap of cauliflower or rather heads of Savoy piled and melted together; with many holes and cavernosities. On the tops of some grow small-sized trees, or an *Opuntia*. At first it seems incredible that a tree should grow upon such lumps of sun-heated lava, so hot on the flatter parts that they are painful to touch, but upon investigation it will be found that the cracks or holes are filled with debris, dusty ashes or sand, and some humus from decayed ferns and lichens. Now, if those blocks exhaled hot vapours, they would be dissolved into earthy matter in the same way as Schleiden described the lava blocks to be acted upon by the fumaroles near the rim of the crater (cf. Schleiden, p. 95), and I quite agree with him that these blocks on the Malpais are the cores of Humboldt's hornitos, but with the reservation that I look upon them as rather potential hornitos. Whether they were originally covered with layers of sand and ashes is a detail, but not every lump rising from the lava field need have acted as a vent for the hot steam. Only those which did were of course hornitos and were, by the action of the steam, converted into the curiously disintegrated hornitos described by Humboldt; and it stands to reason that he selected the most typical. On the other hand, those which did not act as fumaroles, or only for a short time, remained as they were, and if they were originally covered with sand or ashes, these have been washed off and form the deep masses of shifting sand which are still bare of vegetation and fill nearly all the depressions in the Malpais. One of its characteristics is that scarcely any sand lies upon its irregular, uncompromising surface.

Lastly. How have these hornitos, potential or actual, arisen? Humboldt prefers calling them elevation- instead of eruption-cones, clearly because his mind was bent upon elevation. It is indeed a storm in a tea-cup. If anyone had described the scene as a field of lava, of several square miles, on the rough, irregular, terraced surface of which had puckered up ever so many cones, lumps or blocks like so many molehills in a field, that would probably be a description which nobody would have found fault with. Not unlikely some of them did arise like blisters or piled themselves up like ice-hummocks (and there are also huge masses upon the Malpais, far too large and too high for anyone to think of as hornitos) and others may have built themselves up as miniature craters. Humboldt's objection that they did not contain lumps of the older syenitic rocks cannot be of much weight; whilst such lumps of syenite were fired out of the big craters, which brought them up from the bowels of the earth, the stuff which by its overflow formed the Malpais was well-cooked lava, in which any lump of older rock had already been dissolved.

Chapter II

THE PROCESS AND PROGRESS OF RECLAMATION OF THE DEVASTATED DISTRICT

(a) PLANTS

THE devastation was absolute over the whole extent of the volcanoes themselves and of the Malpais, the latter being a space of about $3\frac{1}{2}$ square miles covered with molten lava. The terrain beyond the Malpais suffered of course only from the accumulation of sand and ashes, the severity of damage lessening with the distance. The whole plain to the north and partly to the west, ever since called 'La Playa', the *beach*, was covered to a depth of many feet. At La Huacana (6 miles from the main volcano, which henceforth we shall take as the centre in counting distances) the fall was severe enough to break down the roofs of the buildings; the cattle farm of San Pedro, about 10 miles to the south-west, suffered much, and even at far-off Oropeo the fall of ashes was heavy enough to destroy the pasture. Burkart's testimony is valuable for the condition of things found by him in the year 1827: 'In the neighbourhood of the Rancho Cayaco the ground begins to be covered sometimes several feet deep with ashes from the Jorullo, still 6 leagues (15 miles) distant. Northwards from this Ranch the ashes cover *all* the rocky ground, and only near La Joya and a little further north, grey basaltic rock crops up'.

Similar falls happened in other directions, according to the wind. The seriousness of destruction surely must have presented various degrees. Whilst at some exposed places the trees were laid low, or at least their branches were broken, the damage must have been far less in sheltered spots. We can easily understand that the cattle were starving, as reported, because the grass and the leaves were shrivelled up and covered with the fine dust which made them inedible; but it is impossible to say how far these trees and shrubs were killed or damaged beyond the chance of recovery. However, to judge from the configuration of the ground and from the present distribution of volcanic sands we may assume that plants and animal life were smothered to the following extent: 2 miles to the east and uphill to a height of 5000 feet above sea level; 5 miles to the west, but only 4 miles in the north-western quadrant, the limit being formed by the range of hills; and 2 to 3 miles to the north and to the south. In all, an area of at least 26 square miles, assuming in conformity with the terrain a somewhat egg-shaped area instead of a rough-and-ready oblong of 7 by 5 miles.

According to Ordoñez the whole grand amphitheatre has a diameter of at least 14 kilometres, therefore an area of more than 50 square miles. My estimate of 26 square miles is therefore well within the probability of really serious destruction. Absolute death, however, must have extended over the Malpais and around the base of the chain of volcanoes, a little to the north, east and south, within the range of the bombs and other hot ejections.

The first parts reclaimed by vegetation were naturally the plains of La Playa, these being well watered by streams from the northern slopes. Volcanic sands are notoriously fertile owing to the richness in potash and other salts. Landivar tells us that the fields were unfit to be cultivated for a 'lustrum', a poetical way of saying that years passed before the attempt was made. It has to be borne in mind that the date of the great eruptions which caused the Malpais is unfortunately not known. The birth of the volcano on September 29th, 1759, marks only the beginning of at least four years of greatest activity. Clavigero, whose work was published in 1780, got Governor Bustamante's report about the volcanoes in the year 1766, i.e. about a 'lustrum' after the first outburst. We may reasonably suppose that the people waited until the volcano had ceased to frighten and threaten them by its frequent activity, until at least the year 1774. It is significant that Espelde, a Spanish gentleman, not a native, made the first ascent in 1780. Obviously he reconnoitred the district, and he settled at, perhaps founded, the present La Playa. Nine years later he joined Colonel Riaño, the military commander of Michoacan, and the geologist Fischer in their ascent. The latter tells us that the 'chief devastation had taken place within a circumference of one to one and a half German miles [say about 4 to 6 English miles] which cannot be entered without shuddering'. The Malpais was still uncomfortably hot in places; cigars could be lighted by poking them into an hornito, but it was already covered through its whole extent by established tracks leading to the mines of Inguaran.

By the year 1803, the time of Humboldt's visit, the Playa had become a flourishing place. 'Its pretty vegetation of masses of *Salvia*, blooming beneath the shade of a new kind of fan-palm (*Corypha pumos*), and a new kind of alder (*Alnus jorullensis*), contrasts with the barren, plantless appearance of the Malpais', this being still too warm to support vegetation. In Humboldt's map the sand-covered, absolutely level, plain to the west of the Malpais is inscribed as being under cultivation of indigo, just as is now the case. All the volcanoes were covered completely with grey-white volcanic mud, including the 'white cone of ashes of the main volcano'.

Humboldt, however, makes a surprising statement about the Cerro Partido. 'Only part of it is covered with dense sand. The cropping-up basaltic cliff, *upon which are growing extremely old stems of* Ficus indica *and* Psidium, is certainly to be considered as pre-existing to the catastrophe'. The presence of these trees cannot well be doubted, although it is surprising that such should already have grown up on the rather barren top. Their presence shows that this being old ground, from which the sand had already been removed, was the first and only spot upon which seeds could grow. But the implied suggestion that the 'uralte Staemme' were survivors of the catastrophe, must be dismissed as quite impossible. The basaltic rock is most exposed at the eastern corner of this hill's back, within 800 metres, or just half a mile, from the main crater and only 100 feet above what at one time was a huge field of flowing lava. No tree could have survived in such a position, unprotected from the volcanic eruptions. Further on I shall have occasion to describe how fig trees can assume the look of 'uralte Staemme' within less than 40 years of growth.

By 1827, the year of Burkart's visit, the Malpais itself was being reclaimed by Nature. Most of the hornitos had crumbled or been washed down, 'owing to the very strong rains and to the vegetation which is spreading more and more every day' as Burkart remarked.

Great progress was made during the next 20 years. Schleiden, who visited Jorullo in February, 1846, has given us a sketch of the volcano as seen from the Hacienda Tejamanil, amidst its grove of palms, cypresses, pines, figs and acacias, which seem to extend a fair way down over the plain. 'The upper part of the volcano, consisting of black lava, and all the lava streams, except the first flow which covers the western half of the Malpais, are still without vegetation, but the sandy slope of the main cone, and the Malpais[1] show already considerable growth [sind ziemlich bewachsen]. Some kind of mimosa, not very high, and guava trees are the most important; a pine near the summit seems to me the most interesting of the members of this vegetation.' These are important statements, showing that the vegetation was successfully spreading over the western part of the Malpais, and over the slope of the main cone; the latter, because it stands so near the eastern foothills and those of the Mata de Platano, helped also by the Cerro del Bonete. The bulk of the lava cake, however, was still bare, or nearly so.

Henceforth the growth made rapid progress, as attested by Pieschel's excellent description of the scenery, as he found it in the year 1853 (see p. 96).

[1] Schleiden restricts the Malpais, in conformity with Mexican usage, to that part where the basalt, or lava, appears on the surface, and makes cultivation or pasture impossible, owing to its roughness. Such a place may be full of vegetation of a kind, but it is *mal pais* or *pedregal*, rough, rocky land, in opposition to sand-covered, although perhaps barren, parts.

Thirty years later, 1886, Leclerq is struck with the luxuriant vegetation through which his party made their way up to the crater, near which he noted the guava shrub and various kinds of acacia.

In the year 1888 Felix and Lenk took the first photograph, which shows the conditions as seen from across the palm-studded plain practically as they are now. Unfortunately they did not take much notice of animals and plants, except the passing note of a settlement of ferns near a steam-exhaling cleft somewhere at the rim of the crater. A photographic view of the great breach of the crater shows this to be still without vegetation.

Ordoñez, *à propos* of the Malpais, remarks that the guava trees are plentiful on the Higuera plateau, at the foot of the cones, and towards the east, and that many of these trees are so big as to seem old enough to have been witnesses of the eruption! This may be doubted for obvious reasons. Further, he states quite correctly that the guava trees grow up to the edge of the Malpais, but never on the tufa which covers the lava; there grow other kinds, for instance copal, tepehuaji, higuera, tepamo, cicuillo, all belonging to the tropical zone, whence they have ascended.

Herewith ends the historical survey. Practically, the flora had reclaimed the lost ground by the middle of last century, say within 90 years of the catastrophe. During the last 50 years some of these trees have grown to noble proportions; they have seeded and new specimens have sprung up and spread, and so have no doubt many other kinds of plants which are there now. The process seems slow, especially in comparison with the rapid growth at Krakatoa, but the centre of Michoacan does not enjoy a tropical rainfall. Morelia, the capital of this State and situated on the plateau, has an average fall of 683 mm. (26·9 inches); and my friends at the mouth of the Balsas estimated their yearly allowance—with a period of drought from November to May—at not more than 100 cm., less than half that of Batavia, which has no rainless months.

After this historical survey, which gives a fairly adequate idea of the advance of the vegetation, I proceed to a more detailed description of its present condition, interspersed with observations which may throw some light upon the mode of spreading.

Every available stretch of ground, always excepting the Malpais in its old extent, is under some kind of cultivation; and this implies a considerable amount of labour, such as clearing the ground of weeds and shrubs, thereby keeping large stretches of the terrain under artificial conditions.

The whole of the northern plain is well watered, therefore the richest part of the district; the water drains exactly towards the north-western corner of the Malpais. On the moistest ground are groves of bananas and plantations of sugar-cane; fields of capsicum, watermelons and Indian corn are separated by barbed wire fences from the drier parts which serve for the grazing of cattle and horses. Only the knolls of the little hillocks, deeply covered with loose sand, appeared bare at the end of May, but even these parts were planted with maize shortly before the rains were expected. It is just too high above the level of the sea for cocos palms to yield a reliable harvest.

To the south this plain is separated from the Malpais by very uneven ground, there being for instance the Cerrito de la Cruz, with a large wooden cross on the top, and the Cerro del Paso Hondo, so called because of the deep gorge near it. Consequently in this undisturbed region grows up a perfect tangle of vegetation, mainly composed of acacias and mimosas, which extends right up to and partly on to the Malpais, one of these inroads leading in the direction from the Paso Hondo straight towards the main volcano. To the east of the volcanoes the terrain is quite bewildering, for the foothills of the great southern slope of the plateau come right down to them, except where two little valleys, the Alberca Grande and Alberca Chica intervene. These valleys are perfectly level, being deeply filled in with ashes and sands, underneath which, at an unknown depth, runs water. This, by capillary attraction rises to near the surface. Thus it is that these Albercas are famous for their water-melons.

On the tops and gentle slopes of the numerous knolls are ploughed fields for Indian corn; in between are clumps and clusters of trees, a mixture of tropical and temperate kinds, the latter being mainly oaks, e.g. the roble, which, as elsewhere in this district, was still in its red buds, the leaves not beginning to unfold until the rains came on in June; there was also the encino blanco, already with large leaves, and showing its chestnut-like inflorescence. Here and there is a palo de yugo, the hard wood of which is used for yokes. These trees, still without leaves, and in a blaze of hundreds of large pink-mauve flowers, looked at a distance like almond trees in full bloom.

A little higher up, the slopes become gentler and broader, all under cultivation of maize, with scarcely any trees and shrubs left standing. On waste spots flourishes a composite plant, like a tall shrub. All this terrain is made up of debris from the higher mountains, with big and small water-worn stones, overlaid with sand and ashes to a depth of more than 5 feet, as is indicated by the path we are following, which has worn itself down to the red loamy soil so characteristic of the slopes of Mexican volcanoes. This then was the surface before the eruption. The active cultivation, setting fire to the shrubs preparatory to ploughing, the planting of maize, and the frequent hoeing have produced a broad belt of ground where scarcely anything is allowed to grow which is not wanted, and it acts as an effective barrier also to animals, which otherwise might have crossed towards the volcano from the south-east.

Ascending higher, we come at 4500 feet elevation to the pine forest, composed chiefly of a pine much like our Scotch fir and with similar short cones; another species has long needles and large cones. There are also oaks and a few specimens of *Taxodium*. This pine forest extends up into the high Sierra and looks as if it had never been disturbed, the ground being thickly covered with humus in which grow many sorts of bulbous plants: the lovely *Bessera elegans*, the 'huele de noche', so called on account of its strong nocturnal scent; there are also thick and hairy-leaved begonias, *Tradescantia* with large white flowers, maidenhair and bracken in profusion. Chasms, sometimes 20 feet in depth, have been washed out of the red-brown loam, causing the fall of many a tall tree, and yet none of the pines are really old, say 90 years of age, whilst a mile farther on we find many which are dying from old age; and the same applies to the oaks and arbutus, sufficiently indicating that these lower slopes have been but recently reclaimed by the forest. This is made quite clear if we trace this vegetation down the slope of the plateau to the foot of the volcano and then up its cone. Its whole eastern side is densely wooded. Up to within a few hundred feet on the north-eastern slope is a belt, a little temperate forest of its own, composed of white oak and hundreds of well-grown pines, of the large-coned and long-leaved kind; they are well-grown trees, with seedlings between them, and certainly none of the trees are more than 50 years old. Sr Castrejón, who visited us at the Mata de Platano, could give valuable information about them. According to him, in the year 1872 the vegetation of the volcano reached scarcely half-way up and was then very thin, whilst the rest was still barren and pines were certainly not in existence on the north-eastern slope[1]; but pines had sprung up on the east and south aspects, all of which have been used up since. This information was pleasing, as only a few days before I had taken some

[1] Castrejón's emphatic statement, that in the year 1872 the vegetation extended scarcely half-way up, and even then was very thin, seems to disagree with Pieschel's description, who in the year 1853 found small trees and shrubs almost up to the top. I mentioned this to my friend, but soon found that he was a man who knew his ground. 'Yes, my description refers to what the volcano looked like to me the many times that I have ridden past it on my way to Ario. Forty years ago it did look as barren as it now looks beautifully wooded. What growth there was, must have been small and confined to the deep rills between the ribs, which as you can see from this house are still mostly bare. In the winter, when the deciduous leaves are down, the ribs stand out most prominently, looking quite grey, contrasting with the darker, densely wooded portions.'

trouble in calculating the age of the pine trees. Their seeds must have been blown on to the volcano from the slopes to the east, the nearest spot where these trees were at home; the prevailing winds are easterly, and storms come invariably from this quarter. Wind-blown seeds may often have alighted on the volcano long before it was ready to support a little pine forest, and Schleiden's 'Fichte nahe dem Gipfel' thus finds its natural explanation. The lower limit of these conifers in Michoacan lies near 900 metres. At this level they reappear on the Sierra Madre del Sur, although outposts of the long-leaved kind may, near the Pacific coast, be met with as low down as 800 metres or 2600 feet.

Due south of the volcano lies Mata de Platano, the 'Plantain Grove', not the 'Kill-plantain' as has been suggested by one of the earlier travellers, who thought that the hot springs there prevented the growth of these plants! There are springs, or rather water from the slopes is permanently oozing out beneath the surface, converting the steep southern edge into an almost boggy place.

The little plateau of the Mata is divided from that of the Higuera by a deep chasm, the western wall of which rises almost vertically in many parts and is densely wooded, thanks to moisture which is oozing out of the basaltic plateau. The brook at the bottom forms a little waterfall over a ledge of the old underlying non-basaltic rock. Gradually the chasm widens into a valley with the most luxuriant vegetation, of which the hincha huevos tree is the most remarkable. Here are acacias, mimosas, the cruzada, guava, the uva cimarron, or wild vine, with its appetising-looking pink grapes, nice to taste but leaving a burning sensation upon the lips. Bignonias, with large pods, festoon the trees. This kind of vegetation has made its way up the chasm towards the Cerro del Bonete. In the depression between this and the Mata the black ashes lie at least 100 feet thick, and into these the stream has cut its channel, here and there with banks 10, 20 and many more feet high. Every spate carries down masses of sand. Two streamlets join here: one from the north-east; the other has a peculiar origin, as it can be traced upwards into the sands of the Albercas, and in these sands a brook which comes from the north-eastern foothills loses itself. But more water flows towards the Albercas than comes out of them; it is therefore highly probable that, in addition to other invisible courses, much water is running beneath the lava of the Malpais, in fact along the old bottom, a condition which we shall have occasion to discuss elsewhere.

The Higuera plateau resembles the Malpais in various respects. It ends abruptly in the east and south, and forms several terraces, all inclined towards the south; its highest portion is marked by a series of cones of volcanic origin, besides the Geña Blanca, a beautifully shaped volcano, breached like a volcancito, and called 'Blanca' on account of the pale yellowish colour of the coarse lapilli which cover its upper third. There is but scanty vegetation on the top, only a few trees with sparse grass, and no underwood. It is in just the same condition as the Volcancito del Norte, although the latter is so much younger. Both, being apparently composed of nothing but piled ashes of coarse grain, suffer from drought.

The northern edge of the Higuera plateau, where it

is separated from the Malpais by the series of cones just mentioned, forms a rather intricate terrain and this is very well wooded. Some of the biggest parrotas grow here, and hence their offspring has gained access to the very foot of the volcancitos and to the main cone itself. The central part of the plateau is under cultivation and only a few parrotas, amates and palms are allowed to grow; but wherever the ground is too rough, especially at the abrupt edges of the terraces, the vegetation is abundant, most of all in a valley to the south-west, where a little intermittent stream forms pools amongst the rocks. Here, in sheltered nooks, stand some gigantic specimens of *Cereus thurberi*. Then follow fields, even meadows with cattle ranchos, and then comes the last and lowest terrace with its extensive forest of palms. Only at a few places is it possible to descend into the broad, tropical, luxuriant valley of the Haciendas del Zapote and San Pedro, which lie nearly 1200 feet below Mata de Platano.

To the west of the Malpais, between it and the Canoas range, is a plain, a continuation of the Playa; but whilst the latter is well watered and green, this western plain is quite dry. It is completely overlaid with blackish sand or ashes, to a depth of 5 or 6 feet, as shown by some channels which the rains have washed out. It is so level that it looks like the levelled bottom of a temporary lake, just as Sáyago describes the scene of inundation between the old Jorullo farm and La Huacana.

Humboldt found this plain under cultivation of indigo. At the present time the only vegetation consists of the low shrubs of indigo, here and there with a few palms and fig trees; grass is scanty. The indigo plant is a kind of perennial vetch, a low grey-green shrub with pods which are curled up like caterpillars or snails. Formerly it was much cultivated, but this industry was ruined, when some 20 or 30 years ago a practical process was developed for producing indigo synthetically on a commercial scale.

Not far from our camp, and near the stream, were the now deserted vats—deep, cemented tanks built into the ground, where the indigo had been prepared. The industry is not quite dead in places; there was a rumour that it paid again, just hopeful enough not to burn the whole field of many acres and to sow it with Indian corn, but to wait and see what may happen some day.

Curiosity induced me and the old lieutenant of our escort to try our hand at making 'anil', i.e. indigo. An armful of new plants was chopped up fine and put into the largest available basins. By the next day, after much changing of water and rinsing, there was a nasty-looking, green-blue soup, mostly chlorophyll and woody fibre, which a wondering native laughingly told us to throw away and then to start properly. The branches, with the leaves, are simply submerged in water and left in it for perhaps half a day. Then they are taken out and the water looks as clear as it did before. Then it is to be 'agitated', a boy stirring it for hours with a clean broom, and gradually the water, when in a white basin, assumes a faint blue tint, owing to the appearance of tiny, at first almost imperceptibly small, flakes of blue. It is then left to settle, and if all goes well, on the following morning the bottom of the basin is coated with a thin film of beautiful genuine indigo. When done in earnest, there follow several washings, the sediments being run from one vat into another, until it is pure enough to be dried and rolled into little balls. The slightest addition of lime in the water is said to prevent the natural extraction of the precious colouring matter from the plant.

The palms and fig trees deserve a special chapter. The only kind of palm which occurs at all in the Jorullo district and in the whole depression to the north of the Sierra Madre is the *Copernicia pumos*, first called *Corypha* by Bonpland, who discovered it with Humboldt on the slopes above La Playa. Locally it is known as the *palma real*. There is a tradition that this palm has made its appearance since the eruption of the Jorullo, and certainly its distribution is limited to the extent of the volcanic sand. To the north and east it ends in the neighbourhood of Higueran; to the south-west we saw the last specimens before reaching Oropeo and we did not meet it again when crossing the western half of the depression when on our way to Apatzingan. The corporal of our escort recognised it at once as the *palma di Colina*, he being a native of south-western Michoacan. Don Luis Castrejón of the Zapoti farm, to the south of the Higuera plateau, told me that these palms completely covered his present cattle-grazing ground when his father bought it 70 years ago. They were cut down as interfering with agriculture and pasture and thus, he thought, they were kept in check; but when the people found that the fronds were of commercial value the palms, though not exactly encouraged, were allowed to spread, and as they love sandy soil, at least for germinating, they spread all the more easily and rapidly after the volcanic eruption. Thus he explained the origin of their fabulous sudden appearance. The old fronds are used for thatching and the making of wattled walls. Only the young fronds are valuable. Just before they spread out, they are cut off with billhooks attached to long poles; then they are torn into long narrow ribbons which are then tied together into endlessly long strings which are festooned under the shade of trees to be properly dried, whereupon they are exported for plaiting baskets and especially the universal sombreros, including not a few of so-called Panama hats. The fibres are indeed very pliable, far more than those of most other palms; and ropes, plaited or spun, fetch a fair price.

At first the young tree grows slowly; it does not begin to show a stem until it is at least five years old. Then it shoots up; the ring-like marks left on the stem by the decaying leaves are no indication of its age. Trees 20 feet high are supposed to be as many years old; a height of 40 feet is rare. By that time the base of the stem has thickened very much and adventitious shoots are liable to appear, although these rarely produce a cluster of stems. Excepting man, and maybe some insects, these palms have no direct enemies. Woodpeckers and small parrakeets chisel their nests into the stems, but the enemy which ultimately murders many a palm is a kind of fig tree. This takes place in a roundabout fashion. The seeds of an orchid find their way into the axillae of the decayed leaves, and there grows in time a band of orchids around the stem, a band perhaps half a foot in width and an inch or two thick, formed by the tangled, felted mass of rhizomes and stems of the orchid. This belt, holding moisture, accumulated humus and dust, becomes the inviting nursery of various kinds of other plants. On a few occasions we have seen an *Opuntia* growing in such a belt; even the very palm nuts had been caught in it and were sprouting as young palmlets from the mother stem.

The first and lowest belt stands generally about 8 feet above the ground. Then, as the stem is growing taller, and frond after frond falls off, the first orchid belt loses the shade, it withers and decays, and a new belt is formed higher up nearer the base of the live fronds, mostly some 20 feet above the ground. There is always a long distance between these orchid belts, and never have we found them in close procession, never more than three, usually only two; and tall palms had only one, the topmost, the others having dropped off years before. Why this should be so, I cannot explain. It would seem more natural that the orchid belt itself should creep upwards, by growing at its upper whilst decaying at the lower margin. Further, in these belts grows a fig, always one only! This sprouts out in a slanting upward direction and sends long roots down the stem, or the roots dangle in the air until they anchor in the ground. Within a few years these roots surround the trunk of the palm with a fusing network and ultimately there stands a buttressed fig tree, two yards or more in diameter, with widespread limbs; and above and out of the very centre of the dense, dark green foliage of the fig tree rises the crown of the old palm, a wonderful sight. In time the fig tree overtops even this remnant of the host, and the palm is smothered. The fig tree must exert much pressure upon the palm, but being devoid of bark it lives on for years until at last it decays, mainly from want of light, until the mighty higuera alone remains. This in turn soon acts as host to several kinds of *Loranthus* of the mistletoe tribe, which grow into bunches a yard across and have a peculiar, vividly orange-red inflorescence, so that the whole looks like a bush on fire. The base of the *Loranthus*, where the root ought to be, broadens out and not so much grows into the supporting branch as this in turn grows around it. When dried, the two can be separated by a strong blow and the two wooden star-shaped rosettes, the positive and the negative so to speak, are sold as *flor de madera*, wood-flower—well-known curiosities exported from Spanish American countries, turned into paper-weights and similar mementoes.

These *Loranthi* grow upon many kinds of trees in Mexico. The favourite hosts seem to be the various sorts of fig trees, in the dark green foliage of which the fiery bunches look very handsome, but most striking on the *Amate prieta*, the black fig tree, when this has shed its leaves and before the new have sprouted as they did at Jorullo in the month of June. Other kinds of *Loranthus*, in the temperate zone, grow upon the cotton trees, poplars and willows; others again, right up to the upper tree line, were living upon pines and looked very much like our mistletoe, *Viscum album*. But the most uncanny that we saw were on the Tarahumare Sierra, west of Chihuahua, where they grow upon the cypresses, and the whole configuration of those parasitic bushes bears a really striking resemblance in shape and colour to the sombre grey-green, small-leaved, wind-worried branches of the cypresses themselves. Imagine a cross between a tall juniper and a *Taxodium*, weather-beaten, covered with grey lichen (*Usnea*), which dangles down from the branches, most of which look dead, with a bunch of their flattened, needle-like leaves only towards the tips. On closer inspection some of the best flourishing branches are found to belong not to the tree but to its parasite! Why this appalling resemblance between host and lodger? It would be mimicry, if one of them derived the slightest benefit from the other by their likeness. The only reasonable explanation is that identical environmental conditions have shaped them both alike —the same struggle amidst the same winds, drought and clammy fogs—and yet, within a few yards of such a sight, there grows upon a pine another *Loranthus* with the normal leathery leaves of a mistletoe!

To return to the climbing fig trees. Sometimes if a fig has settled upon a palm which was still too young, and before the fig itself has sent sufficiently strong buttresses into the ground to support its own weight, it is blown over and it bears its host down. The interesting question is, how did the fig get on to the palm stem, usually some 5 feet above the ground? As my wife put it, the sapling figs looked as if they had jumped into the nearest stem available. I think we have found the agents—bats. The masses of dead fronds beneath the crown of the palms are a favourite retreat of bats, and there is, in southern Mexico, a small kind of fruit-eating bat, *Glossophaga soricina*, which is addicted to the ripe globular little figs of the higuera, which are only the size of a cherry. The roof of the house at Mata de Platano was tenanted by these bats, and every morning we found their dung and half-eaten debris of these little figs on the verandah, whither they had carried the fruit from an old tree which stands some 150 yards from the house. If bats take the figs to a palm in which they lodge, to eat them there, some of the tiny seeds are sure to fall into the orchid belt and the whole process of distribution is beautifully clear; only one does not understand why there is never more than one fig tree sprouting from each belt.

Not all the kinds of *Ficus* start life in this curious fashion; some sprout from the ground, but it is certainly advantageous to the epiphytic kinds to begin well out of reach of harm, since the young leaves are eagerly eaten by cattle and perhaps by other animals. Within the reach of cattle all the big trees are as neatly trimmed from below as if they had been clipped with shears.

The plain to the west of our camp was deeply covered with volcanic sand and studded here and there with palms. These grow just as well between the rocks, provided there is sand. The Higuera plateau, on the other hand, supports a regular forest of palms although there is but little sand, or rather most of the black ashes have been washed off the long gentle slope. The ground is covered with light yellow 'toba' tuff, the fine-grained and denser sediment of the water-washed ashes. No palms grow on the Malpais, although they come close to its edge on the west and south, but—as if to show what can be done by way of exception—two or three specimens are growing on the very top of the western edge of the crater! How did they get there? It is out of the question to resort to storms, which scatter the little nuts far from the parent tree and thus are the means of rapid distribution. Nor can rodents be held responsible. The only practical suggestion is macaws, notorious carriers of even heavy palm nuts, which we have seen flying actually right across the volcano. A pair of these gorgeous birds used to come up every morning from below to spend the day in the crowns of certain big parrota trees.

My friend Castrejón felt sure that the palms on his ground (cf. p. 36) were not killed by the 'revolucion del Volcan', as they could not have become so plentiful and have spread over so large a space within 70 years after the eruption. But he thought it most unlikely that any could have survived on the western plain, between the Malpais and the Canoas range, where they must have been exposed to the full fury of the elements. You may scorch a stem, and new shoots will sprout from the roots; you may bury it to the extent of many feet; but there is one thing that a palm cannot stand: you must not lop off all its fronds at one time, since this suffocates it and it dies down to the ground. This is precisely what must have happened through the long-continued and repeated falls of the volcanic ashes, and this at a time of year when the new fronds were in their prime condition. However, one thing is quite clear, that none of the present trees witnessed the catastrophe, and that they have greatly increased their range owing to the deposition of those sands which afford their seeds a favourable ground for germination. In this respect the palms and the birth of the volcano stand in direct correlation. They in turn have invited the orchids which have prepared a breeding-place for the fig trees, with bats as carrying agents, and lastly under the shelter and shade of these rapidly growing giants whole colonies of underwood-making shrubs are settling and thus converting the once absolutely lifeless plain into a park-like savannah. When these figs are absent, for instance in the palm forest proper, although many of the trees are belted with orchids, the ground is remarkably devoid of anything like underwood, almost bare during the dry season; but grasses and a thick carpet of *Tradescantia* sprout up within a week of the first heavy rains.

A most instructive view of the present state of the Malpais is that from the top of the Veladero, an old cone rising several hundred feet above the neighbourhood. Its own lava flow is the Pedregal which extends towards La Puerta and is now well under cultivation of maize, unless kept as grazing-ground. Its natural vegetation has therefore not been allowed to make much progress since Pieschel's visit in 1853; on the contrary, it has been checked. The name of Veladero, 'the looking-out place', is well chosen. The scene is grand and so beautiful that one fails to see why those several square miles of park-like landscape could ever have been branded Malpais. Its south-eastern quarter appears densely wooded and this forest extends right up towards Mata de Platano, curving round to embrace the row of volcanoes and the little Cerro Partido. It is an apparently continuous mass of green leafy tree tops, and our delighted gaze is held again and again by the white cone of the Volcancito del Norte, which still seems to stem the flow of jet black lava tumbling from the big crater. Towards the north the vegetation is more scanty and there are wide patches of sand, black or pale as the case may be. It is everything but a plain, there are so many dales, terraces, gorges and knolls. There is also water, at least under the surface, the presence of which is marked by the brighter green and denser belts of fine trees with much underwood, which extend from the western abrupt edge on to the Malpais. One of these 'arroyos' or streamlets emerges a little to the south of La Puerta, probably the cascade sketched by Humboldt as falling over the barren precipice. Now all this is a tropical jungle of fine parrotas, grand guavas upon the leaves of which feed the large vegetarian iguanas; and there are bamboos, and festooning creepers of vines and bignonias, the home of the 'salamacoa' or *Boa imperator*. The rushing stream with the cascade has, since Humboldt's visit, dug itself a deep gorge. After the lava crust on the top had broken down its work was easy in the many layers of ashes and loam, the mud-flows of the early days of the catastrophe, until the now tiny brook reached the denser original bottom. It can be traced only a little way on to the lava flow. Another much larger brook falls over the extreme south-western corner, above the few houses of Agua Blanca; its water has still the reputation of being tepid, at least it is said not to feel chilly in the early morning and people are known to have sat in its pools. It passes along the northern side of the Pilas hills, whence it may be followed upwards through the deep Barranca del Guayabal to the very foot of the Volcancito del Sur. Connected with it is the spot known as Agua Ascondida, the 'hidden water', a series of shallow pools in a depression half a mile to the north of the Barranca. These waters are permanent, and interesting because they do not merely collect rain or surface water, but ooze out from beneath the lava field, which they enter at the Alberca on the north-eastern side of the main volcano.

Schleiden was the first to recognise this. 'The waters of this brook [from the eastern foothills] reappear a few miles further west as those springs which, although now cooling from year to year, are probably still being heated beneath the lava.' The different levels—Alberca more than 1000 metres, western foot of main volcano 930 metres, Agua Ascondida 860 metres, Guayabal Barranca 800 metres—agree well with this supposition. In fact we can by this mostly hidden stream reconstruct the bottom of the sloping plain of the original Jorullo paradise. It needs no further comment that all these various brooks and pools have greatly facilitated and given a lead to the colonisation of the new terrain.

Concerning the rest of the Malpais—Pedregal or 'Stony ground' is that portion which for some reason or other is no longer covered with sands. It supports a better vegetation than the flatter stretches nearer the volcano, but all the trees and shrubs are small and there is scanty grass. There are only acacias, mimosas and some copal, very few fig trees and a few *Opuntia*. Copal and some other shrubs, or an *Opuntia*, are growing out of the very tops of the hummocks of lava; maidenhair and a plant like *Selaginella* are found in every crack, shrivelled and apparently dead, but sprouting with the first small showers of rain. That the shrubs and trees are so small is merely the result of the soil; the lava is still too young to have filled, by its disintegration, its own clinker-like cracks and holes with that firm, sandy, reddish stuff which fills a really old pedregal and turns it into a luxuriant rock garden. In such a condition is for instance the Pedregal between Tlalpan and Mexico City, and another near Morelia, both of which are many times older than the Jorullo, although so recent that charred remains of human implements, and even casts of the cobs of maize have been found beneath the lava. Still older is the Pedregal of the Tancitaro, near Uruapan, which, favoured by more rainfall, is now like a rock garden of bewildering beauty.

As we proceed eastwards, the pedregal gradually ceases as such, simply because it is still thickly overlaid with black sand, so as to form gentle sloping, here and there slightly rolling, plains. Here the vegetation is park-like: the mimosas, acacias, copals are larger, but they stand further asunder, presumably because they have not yet had sufficient time to claim the ground, much of which is grassy, unless it is still barren. The distribution and behaviour of the trees is very tricky. Although, on the whole, the average age is clearly less, considerably less, than that of those which grow near the southern border and in the south-western quarter, there are already some good-sized parrota trees forming little clumps. Whilst the hard woods, e.g. mimosas and the thorny acacias, seem to have planted themselves with predilection upon slight, sandy knolls, perhaps the remnants of hornitos, and, whilst a red-barked tree grows out from the most uninviting cracks of protruding lava, the parrotas seem to want finer sand and plenty of it. Where this is shallow, they do not grow at all, but where it lies deeper and upon irregular ground, as for instance at the headland of the deep gulleys, there they do well and are found in company. The beds of these sometimes broad gulleys are quite devoid of vegetation, because they are filled with shifting sands. Generally the flatter and more uniform the surface, the scantier are the trees and shrubs, and only few sorts of them; the rougher, more irregular the terrain, the greater are the variety and the individual numbers of the settlers.

The main volcano is well wooded, densely in the curious depressions which the rains have furrowed out, radiating from the top; the intervening ridges or ribs are bare, but even these are being covered, the growth proceeding clearly from the bottom upwards. Naturally, more ashes, sand and debris are found below than near the crater where the vegetation is more scanty. It is strongest on the east and south slopes, the weather side, whilst the upper third of the western slope is almost treeless, but also steeper; it still holds a large amount of pale yellowish lapilli and is mostly covered with *Bromus* and other harsh grasses.

The vegetation is composed chiefly of parrotas, tepehuaji, mimosas, copal, tacoti, palo jiote; on the eastern side, roble and encino, the pines mentioned on p. 31, agaves with narrow, barbed leaves and tall inflorescence, a red-blooming *Salvia*, a few climbing fig trees, *Loranthus*, bracken, maidenhair and a *Selaginella*-like plant grow in profusion.

All this mass of herbs, shrubs and trees thins out towards the crater, the condition of which has already been described, but I have to add that the breach in the north wall is now being fairly colonised by copal and mimosa, whilst this breach was still barren in 1888, to judge from a large-sized photograph taken by Felix and Lenk.

The whole mass of the black, tumbled-down flow of lava is still almost quite bare, no doubt because no sands have ever mitigated its inhospitable surface. But at the foot of this lava torrent the vegetation is now in the interesting stage of creeping on to it. It took me some time to climb up 50 feet over and between the big, often cube-shaped, sharp-cornered blocks of lava, huddled upon each other like the concrete blocks of a breakwater; and yet here and there a copal, a fig, an acacia, an *Opuntia*, a tuft of *Bromus* or one of the leafless Salvias had found a footing in the cracks. All these specimens were still small. The figs had to manage without a nursing tree and climbed like ivy over the boulders, but it seemed ridiculous that a copal and even one of the pungently smelling red-barked shrubs which are generally shunned, still mere saplings, were already supporting a *Loranthus*, likewise in the baby stage. Nests of *Selaginella*-like plants were quite common, and in deeper cracks, sheltered from the sun, maidenhair was sprouting, and plenty of a hard, very thin kind of grey lichen covered the face of the blocks, which in the noonday sun were so hot that I could not hold on to them. Some 50 feet of that climb was quite enough, especially since the vegetation itself did not extend much higher up, except the lichens and *Selaginella*-like plants, which seem to be the pioneers for this kind of rockwork and which by their own decay and catching of dust produce the humus necessary for more highly organised plants.

Chapter III

THE PROCESS AND PROGRESS OF RECLAMATION OF THE DEVASTATED DISTRICT (*cont.*)

(*b*) ANIMALS

ON the devastated terrain of the Jorullo have been found 33 species. In how far can this number be taken as representative of the fauna of that district before the eruption? The whole calculation being precarious it would be unprofitable to speculate too much about 'what might be' by taking the total presumable number of Michoacan as our basis. We must, instead, content ourselves with scrutinising the actual list of the 86 species ascertained for Michoacan, and subtract from it those species, which for obvious physical reasons cannot well be expected near the Jorullo.

Such 'excused' species, whose absence can be properly accounted for, are the following:

1. Those which, to judge from their behaviour in other parts of South-Western Mexico, live at higher altitudes, and are not likely to descend to the level of, say, 1000 metres or 3300 feet. The Playa is about 800 metres, Mata de Platano 1100 metres, above sea level.

Amblystoma tigrinum	*Eumeces lynce*
Spelerpes belli	*Phrynosoma orbiculare*
Bufo compactilis	*Sceloporus torquatus*
Borborocoetes calcitrans	*S. aeneus*
Hylodes rhodopis	*S. scalaris*
Eupemphix sp.	*S. heterolepis*

50 RECLAMATION OF DEVASTATED DISTRICT

Sceloporus microlepidotus
Gerrhonotus imbricatus
Tropidonotus ordinatus
T. scalaris
Contia nasus

Dromicus laureatus
Pituophis deppei
Crotalus triseriatus
Cinosternum hirtipes

21 species.

2. Those which remain well within the Tierra caliente:

Rhinophrynus mexicanus
Phyllomedusa dacnicolor
Ctenosura 5-carinata
Mabuia agilis
Sceloporus siniferus

Cnemidophorus immutabilis
Himantodes gemmistratus
Loxocemus bicolor
Crocodilus americanus
Caiman sclerops

10 species.

Thus, with 33 species observed and about 30 excused, only 23 remain to be accounted for. Of these, however, at least half are so rare, or local, with our present knowledge of their distribution (e.g. *Hyla arenicola, Coleonyx, Geophis dugèsi, Typhlops*), that scarcely a dozen species remain as probable or possible desiderata for the Jorullo district worth accounting for. These should be interesting.

1. *Hyla copei* would, to judge from its distribution as known elsewhere, find its lowest level at 2500 feet just on the surrounding foothills of the Playa, whither it could have descended from the north and east. This species is not arboreal but prefers rocky ravines with boulders, and little pools of water during the dry season. During the wet months it is often met with on stony, barren ground, but it avoids dry sand. It could scarcely be expected in the district; moreover, its only record for Michoacan is a permanent, rocky stream above Apatzingan.

2. *Hyla eximia*, a small tree frog, prefers moist meadows; as a rule restricted to higher levels, it occurs at Presidio near

Mazatlan, and I have found it as low down as 2000 feet in Guerrero. This species would have no chance in descending from the Great Plateau.

3. *Leptodactylus albilabris* may well be expected, because elsewhere I have found it together with *L. caliquiosus*, which is plentiful in the streams of Jorullo.

4. *Hypopachus variolosus*. Previously recorded from 'Morelia'; found by myself at San Salvador, Buena Vista, and at Cofradia, therefore to be expected in the Jorullo district, where, however, it does not seem to occur in spite of suitable terrain. I suspect the volcanic ashes to have smothered these small, retiring and rather slow creatures which feed to a great extent upon termites, insects which are still absent from the district.

5. *Uta bicarinata* is perhaps a variety of *U. irregularis*. This is apparently still rare at Jorullo, whilst in other similar localities it is easily seen and collected in the shrubs and trees which this rather slow species does not willingly desert.

6. *Phrynosoma asio* is humivagous and slow. I have found it only on rocky and rather well-wooded ground, covered with shrubs, never upon sandy slopes. It does not seem to be known about Jorullo, whilst in the western part of the depression, at Cofradia, it was plentiful; also on the Sierra Madre del Sur, for instance at San Salvador, at precisely the same level as La Playa, but this seems to represent its top level. It seems reasonable to presume that this species has not yet recovered from the effects of the eruption.

7, 8. *Sceloporus scalaris* and *S. aeneus* are both humivagous on grassy places, prefer as a rule the higher altitudes, but in Guerrero I have found them to descend well into the Tierra caliente. These easily observed species were plentiful at San Rafael, 8000 feet, south of Patzcuaro, and on the upper, northern slopes of the Jorullo depression, whence they might easily have descended below Ario, towards the Jorullo. But they have not done this, perhaps on account of the volcanic sand, although on Citlaltépetl they flourish upon the same kind of sand or ashes, almost up to the snow line, but of course in a much moister atmosphere. Besides at San Rafael, I found them in Michoacan only near Aparicuaro, half-way up the Tancitaro mountain. These species may therefore be 'excused' the Jorullo, ranking in Michoacan as decidedly temperate or higher level.

9. *Sceloporus formosus* might well be considered as a possible descendant from the Great Plateau down the northern and eastern slopes of the Jorullo basin. That it actually does descend even further down is shown by its rather unexpected occurrence at Cofradia.

10. *Basiliscus vittatus*. These arboreal and aquatic, herbivorous, very agile lizards require permanent streams with trees and shrubs overhanging. The Zapote and San Pedro valley seems to be a perfect locality for this species, which is sure to attract notice. Although it has been recorded from Parécuaro, I have met with it nowhere in the depression north of the Sierra Madre, whilst at Carrizal and again at the coast, it was plentiful.

11. *Heloderma horridum*. Humivagous and slow, preferring, at least in Michoacan, woods, or fairly wooded localities with soft soil, preferably humus. It ranges from the coast up to 4000 feet, crossing the Sierra Madre into the Balsas depression, being well known at Apatzingan, Parécuaro, and at Oropeo to the south-east of Jorullo; it has been recorded also from Etucuaro, about 40 miles to the north-east. The absence from the Jorullo district of this slowly travelling, much dreaded creature, may be another instance of the effect of the smothering of the country with the erupted ashes.

12. *Glauconia* has been found by such an absolutely reliable authority as Dugès, near La Huacana. These little burrowing snakes may easily survive a smothering with sand, but their records are too scanty.

13. *Tropidonotus ordinatus*, although one of the commonest snakes in Mexico, with the enormous altitudinal range of 10,000 feet, including the coast lands, is an entirely high-level species in the south-west, behaving in this respect like *Sceloporus scalaris*.

14. *Coronella micropholis*. I cannot account for the apparent absence of this snake from the Jorullo district proper. It occurs at the Zapote Hacienda.

15. *Crotalus*. Rattlesnakes seem to be unknown at the Jorullo. They are slow, bad migrants and very local. The Malpais in its present condition, and its vicinity, would afford an ideal ground for these brutes provided they could get there. The broad belt of sand-smothered terrain on the west, north and east seems to have acted as an effective barrier and is now too well cultivated. On the southern side the Higuera and the Copales plateaux are difficult of ascent, and the dense tropical vegetation of the well-watered valley of the Zapote and San Pedro farms is not suitable for rattlesnakes. Therefore we concluded that they have been killed out by the eruption, and have not had time to re-occupy the district—provided that they ever occupied it, as was most likely the case.

To sum up, concerning the 'desiderata' which, *à priori*, might be expected to occur at or near the Jorullo, but which have not been found there: Six species (viz. *Hyla copei*, *Hyla eximia*, *Sceloporus scalaris*, *S. aeneus* and *S. formosus*, and *Tropidonotus ordinatus*), all active snakes, would have had to descend, to reach the Jorullo, which would be rather below their usual level. For nearly all the other species the Jorullo plain would be their proper level to occupy, which would imply nothing but the closing in from the neighbourhood without any change of altitude. It is significant that the majority of these species, certainly *Hypopachus*, *Phrynosoma*, *Heloderma*, *Glauconia*, and *Crotalus*, are sluggish and non-arboreal.

We can now proceed to a scrutiny of the 33 species which are known to occur within the once-devastated district, creatures which, we have good reason for supposing, have immigrated from the neighbourhood. They have been marked as descendant or as ascendant according to the considerations discussed in my papers on 'The distribution of Mexican Amphibians and Reptiles', *Proc. Zool. Soc.* 1905, II, pp. 191–244, and 'The effect of altitude upon the distribution of Mexican Amphibians and Reptiles', *Zoolog. Jahrbücher*, Abth. System. vol. XXIX (1910), pp. 689–714. For instance, *Iguana rhinolophus*, an essentially tropical species with an upper limit of 3000 feet elevation, has clearly to be treated as having ascended from the Tierra caliente— in the present case from the southern valley of the Hacienda San Pedro. A few remarks have been added about the habits of the various species.

Bufo marinus, B. marmoreus, B. valliceps. All these toads require water for spawning and during the tadpole stage. During the rest of the year they are independent and can spread over considerable distances from their native spot. During prolonged drought and heat they aestivate. Although essentially tropical, the Jorullo is well within their altitudinal range, but the configuration of the whole terrain indicates them as ascendants.

Hyla mocquardi, H. bandini. Arboreal, absent in pine forests, requiring water; not aestivating. Ascendant.

Hylodes palmatus. Proper level. Arboreal and partly aquatic, preferring moist, shady underwood.

Leptodactylus caliginosus. Aquatic, in streams, pools, or puddles and springs between the roots of trees. Ascendant, with 3000 feet upper limit.

Rana palmipes. Aquatic. Well within its range, but obviously ascendant.

R. halecina. Aquatic. Proper level. Ascendant of the brooks.

Cinosternum. Aquatic. Ascendant of the brooks.

ANIMALS

Uta irregularis. A very small iguanid, arboreal mostly on shrubs, dry surroundings. Proper level.

Sceloporus spinosus. Arboreal, mostly on shrubs, but also a good runner; dry rocky localities. Proper level.

S. melanorhinus. Same as above. Ascendant.

S. cupreus. Ascendant.

S. pyrrhocephalus. On and between rocks with shrubs. Ascendant.

S. gadoviae. Ascendant.

Cnemidophorus communis, C. deppei. Swift runners on grassy, shrubby, sandy ground.

Iguana rhinolophus. Arboreal and semi-aquatic leaf eater, always in the neighbourhood of water and trees. Ascendant.

Ctenosura acanthinura. Omnivorous, ubiquitous. Fast runner, climber of rocks and trees and good digger. Independent of water.

Anolis nebulosus. Arboreal, also on herbs. Insectivorous. Proper level.

Phyllodactylus tuberculosus. The common gecko of Mexico. Arboreal. Ascendant, or proper level.

Zamenis mexicanus and Z. lineatus. Fast runners and good climbers. Proper level.

Coluber corais. Fast runners, partial to water. Ascendant.

Drymobius margaritiferus. Arboreal. Ascendant.

Oxybelis acuminatus. Arboreal, most frequently on shrubs. Ascendant.

Leptodira personata. Mostly on the ground. Proper level.

Trimorphodon biscutatum. Typical ground snake. Ascendant.

Conophis vittatus. On the ground, and digging.

Elaps fulvius. Ground snake, requires good vegetation and mouldy soil. Proper level, but clearly ascendant.

Boa imperator. Good climber, but mostly on the ground, always amongst good vegetation and within reach of water. Ascendant.

Glauconia albifrons. Tiny burrowing snake; often under stones on grassy, rough ground. Ascendant.

Out of the whole list only the *Iguana* is herbivorous, an eater of leaves and fruit; the fierce *Ctenosura* is omnivorous, eating leaves and even grass, and any kind of creature which it can master. The other lizards are almost entirely insectivorous. All the snakes depend mainly upon lizards, mice, or other snakes, whilst a few, like *Zamenis* and *Coluber*, take also birds.

If we divide these species of Amphibia and Reptiles into the three groups of high-level, proper-level and low-level species, we find that high-level species are not represented, whilst those of the proper level are either dependent upon water (therefore necessarily to be classed as ascendants) or they are good runners, or arboreal. In fact none of the actual fauna has necessarily to be considered as having *de*scended upon the scene. In contrast herewith stand the so-called desiderata; amongst them, the only ones which would have to ascend are the slow *Phrynosoma asio*, *Basiliscus* and *Coronella micropholis*. The species of the proper level, i.e. those which would simply have to close in from the periphery, without altering their level, are also slow and non-arboreal. Lastly, those of the higher level, at least 6 out of 13, are rather agile. All this clearly means that there was no inducement to descend, not even for the rather easily moving kinds, whilst the closing in from the same level has not yet happened, because the respective species are slow travellers. The slow species of the actual fauna all belong, as we have seen, to the ascendants. To put these results in other words: Half the number of desiderata are high-level species, which have not descended although they are fast and independent. No slow or sluggish species has closed in from the surrounding foothills, or, in other words, none of those which are slow to move have as yet had time to close in from the foothills.

The volcanic sands thrown upon the slopes to the west, north, and east of the grand amphitheatre have delayed, or are still preventing, the recuperation which, on the contrary, has proceeded from the south, guided or assisted by the streams. But for them and their concomitant vegetation, 150 years would not have been sufficient for such a rich fauna to have reclaimed the district.

According to the configuration of the whole terrain there is little doubt from which parts the colonisation proceeded. Scarcely any species came from the eastern slopes, except *Sceloporus spinosus*. The gradual invasion began in the rich southern valley of the Haciendas San Pedro and Zapote, following up the streams. In the south-east a long-continued check must have been experienced at Mata de Platano; but in the west the western plain, thanks to the river, was easier to reclaim and led to the rapid settlement of the northern well-watered plain. Thus it came to pass that the Malpais, with its central position, was attacked from three sides, concentrically. There should be, with this assumption, and if the process of reclamation has not yet been completed, a much scarcer fauna on the Malpais and still fewer species in its centre.

Whilst, then, the fauna of the whole amphitheatre is known to be composed of as many as 33 species, we have found only 6 species (*Cnemidophorus gulanis*, *C. deppei*, *Sceloporus pyrrhocephalus*, *Uta irregularis*,

58 RECLAMATION OF DEVASTATED DISTRICT

Ctenosura acanthinura and *Zamenis*) on the Pedregal, to the west of the volcano and around the Cerro Partido; and only 2 species (*Cnemidophorus deppei* and *Ctenosura*) on the top of the volcano.

Of course the richly wooded, intricate, and partly watered southern portion of the original Malpais contains many more species, whilst not many can be expected on the Pedregal and away from the water, for instance, none of the Amphibia and tortoises, nor the *Iguana*, *Boa*, *Coluber* and *Elaps*; we may further take off *Oxybelis* and *Drymobius*, but even with this reduction of the grand total, we have actual evidence of only 6 out of 17 which might reasonably be expected. Of course our not having found certain species does not prove that these do not occur. I can only judge from appearance and experience, from the fact that we found so many more kinds upon the outskirts of the Pedregal than on this place itself. For instance, one may be pretty sure to see the various kinds of *Sceloporus*, the *Anolis* and the gecko, wherever they exist; and not finding any of them on repeated visits is therefore fair enough evidence. Moreover, the native boys, some of whom collected briskly, could not be induced to hunt on the Malpais proper, because they knew that they would return empty-handed, especially after they had learned that specimens of the ubiquitous *Cnemidophorus* were not in request.

Well then, 32 species around the very edge of the Malpais, only 6 on its centre and only 2 on the top of the volcano, is not a bad result, because these rapidly decreasing numbers seem to show how this area has been reclaimed by working inwards, and we may well assume from these data that the recolonisation is not yet finished.

The little, but swift, *Cnemidophorus deppei* has clearly ascended the volcano by its western more grassy and sandy, but steeper slope. The big and strong black iguana, *Ctenosura*, does of course roam everywhere. The big *Cnemidophorus*, a very active runner, cannot be induced to leave rather level ground provided this is covered with a fair amount of shrubs or open underwood. On the other hand, the little *Sceloporus* loves large boulders and on the Malpais is to be found only where the lava appears on the surface, whilst it avoids the levelled-down sandy parts. It goes right up to the basaltic blocks at the very foot of the volcano but not on to it, 2500 feet being the upper limit of this tropical species.

On the top of the volcano, among the tussocks of grass on the western side, was the dung of mice and of hares, even a cow seems to have made its way up there, but besides the small *Cnemidophorus* and the black iguana, and of course insects, we found only black and yellow scorpions and black centipedes, resembling *Julus*. The latter were also common enough on the almost barren tops of the little volcanoes, still more plentiful, however, at their foot—more than sufficient for our purpose. Indeed, far too many scorpions lived at our camp and at Mata de Platano. All the more surprised we were at not finding any scorpions or centipedes at the centre, around the Cerro Partido, although we turned many a stone.

The prime impelling agent in the colonisation of a new area is the search for food. The majority of creatures, in our case the Reptiles and Amphibia, do not roam about from lust of adventure. On the contrary they are much attached to their chosen abode, a hole in the ground or in a tree, a cleft in the rocks, to which they will always return, unless they are crowded out or they happen to find a better one, or one more conveniently placed, during their search for food. This and the want of suitable breeding-places induce them to venture further afield. Elemental forces like wind and the currents of water may be left out of consideration, and moreover, at Jorullo the spreading went up stream. In many cases the distribution of a species is discontinuous, local or patchy, and is sometimes puzzling to account for, seeming to have been effected by jumps. This applies not so much to quick and enduring runners but rather to species of which one would least expect such vagaries, for instance toads. They have, like most Amphibia, to spend their tadpole-stage in water, but after that they are quite independent of it and, since it takes them several years to reach maturity, they have a good chance of getting far away from their native spot. If, during their 'Wanderjahre', they happen upon an isolated pool, they will most likely breed there and found a new centre of dispersal with their numerous offspring. If, on the other hand, there is no such pool, they will return to whence they came, unless their own experience, or may be their 'instinct', leads them to the nearest available spot. In reality the phenomenon is not quite as wonderful as it may appear. It looks as if all the toads for miles around had congregated at a certain pool, but we do not know, and therefore take no cognisance of the countless misses, which were not clever, or not lucky, enough to find that spot.

ANIMALS

With due consideration of the question of food we may assume that the first Jorullo pioneers were the insectivorous and ground-running lizards, which in their turn drew the lizard-eating ground snakes. Later come the arboreal hunted and hunters. The simplicity of this sequence is, however, chequered by the means of dispersal possessed by the various creatures. Whilst a black iguana, or a *Zamenis*, could, if so inclined, cross the whole Jorullo district within a day, this would take a *Sceloporus* or a *Trimorphodon* many days and the little *Uta* would require weeks. A *Cnemidophorus* finds no difficulty, provided there is sufficient insect life, but such typically arboreal snakes as *Oxybelis* would not easily be induced to leave the tangle of shrubs in order to cross on to another cluster; on the contrary, this snake, or perhaps its offspring, would have to wait until these clusters join. The coral snakes, eminently nocturnal and leading a mostly underground life, hunting between moss and mould, would not cross a stretch of barren ground; and a rattlesnake would not willingly ford a stream. The slowest and most precarious to extend their range are the little Glauconias, which almost resemble earthworms in their underground life. We may use them here as a type of creatures which extend their range not so much individually, not by themselves, but by their offspring, rather like plants. The children require more space than the parent pair, and so do the grandchildren, each new pair forming a new centre, so that the spreading proceeds in epicycles and the whole population expands in local continuity. Strictly speaking, and from a broad point of view, most animals and plants are spreading in epicycles, the range of the species being expanded by the offspring at the periphery. Sudden rushes, irruptions and other accidental or occasional dislocations (often misnamed migrations) lead mostly to nothing. This is the rule and is compatible with the well-known fact that currents of air and water, sand and water-spouts, animals, and even the proverbial floating log, can or may help the dispersal of animals and plants.

Lastly, is it possible to calculate the rate of spreading? Unfortunately the only factor known is the utmost time limit, namely, that the fauna and flora, wherever they exist now, have got there within 150 years, but there is no means of ascertaining how long ago a certain animal species may have settled at the spot where we find it now. Trees at least allow the calculation of their age, and the various travellers have mentioned the occurrence of various kinds of plants at different localities, but to the fauna none have paid the slightest attention, except the statement that somebody has seen a fox, a stag or a snake.

Another drawback is the impossibility of ascertaining to what distance from the centre life was completely destroyed. All over the volcanoes, and so far as the lava extended, the destruction was of course absolute; but beyond it, even in its immediate neighbourhood, it was at least possible that certain kinds of animals and plants did survive in sheltered nooks and corners, for instance, under the overhanging walls of the deep gorge to the west of Mata de Platano, or at the south-western corner of the Malpais. However, even if some had survived the catastrophe, the enormous amount of fallen ashes, most of which have since been carried away to the depth of

ANIMALS

several yards, would have rendered life, feeding and breeding practically impossible for years to come.

If we assume that the whole amphitheatre has been reclaimed in less than 150 years and at the very least from a distance (reckoning the middle of the Malpais as the centre) of 2 miles, the rate would be 1 mile in 75 years. This is far too slow an estimate because, first, the recuperation was already fairly universal at Pieschel's visit in 1853, i.e. about 90 years after the catastrophe; this gives a rate of 1 mile in 45 years. The same rate results from the consideration that the Malpais was still barren at Humboldt's visit in 1803, but well on the way to recovery at Pieschel's time; rate 1 mile in 50 years. Further, the Playa had already recovered by 1803: this means that a belt of at least 2, more likely 3 miles, to the north and west from the edge of the Malpais, had recovered in about 40 years, a rate of 1 mile in 13 to 20 years.

The rate of course becomes slower the further we assume the severity of the falls of ashes to have extended. On the other hand, the process of recuperation is hurried on upon a well-watered Playa, and is very much slower on the dry western plain, fit for a few sorts of plants only. Many calculations, all necessarily vague, favour an average rate of 1 mile in 40 years. The general rate cannot have been as slow as 1 mile in 80 years, and only in very favourable spots can it have been as fast as 1 mile in 10 years. Of course these calculations can be applied merely to the general process of recuperation, to the progress made in the general work of improvement of the original devastation. In detail there are, and must be, many exceptions. For instance, Colonel Riaño, as early as the year 1789, already found shrubs of the nettle-tree somewhere up the volcano, which implies that its soft, white, mulberry-like fruits had been carried thither by birds within 20 years after the last bad eruptions, which we are told continued until the year 1766. If Humboldt's 'very old trees of figs and guayabas' on the top of the Cerro Partido are not a myth, their edible seeds must have been carried there at the earliest possible date, and even in that case the trees he saw could not have been 40 years old.

The rate of 1 mile in 40 years may seem slow for a swift lizard; but it seems quite fast enough for scorpions, which also occur on the top of the main volcano, at the very least 1 mile from the nearest possible starting-point and more than 1000 feet above it.

At this rate a square of 4 square miles, the shape and size of the absolute devastation, would, by concentric attack, be covered in 40 to 50 years. If we apply this to the Jorullo volcano and the Malpais, which were barren 100 years ago, it may now well support trees 50 years of age, a condition of things which is fairly borne out by the parrotas, figs, palms and guavas.

The whole process may appear slow in comparison with such a rapid recovery of plant life as that of Krakatoa, but the centre of Michoacan does not enjoy a tropical amount of rainfall. There is a long rainless period from November to the beginning of June. Morelia, the capital of the State, and situated on the plateau at an altitude of 6365 feet, has an average rainfall of 68·3 cm. And friends at the mouth of the Balsas, where it rains heavily when it does, estimated their yearly allowance at little more than 100 cm., less than half that at Batavia, which has no rainless months.

Chapter IV

THE AMPHIBIA AND REPTILIA OF MICHOACAN

IN a paper on 'The Distribution of Mexican Amphibians and Reptiles', *Proc. Zool. Soc.* 1905, II, pp. 191–244, I made the following remark: "The whole State of Michoacan and the western half of Guerrero are still an almost absolute terra incognita, but to judge from what I have found in Middle Guerrero, from Iguala to the coast, and what is known from Colima, the fauna seems to be rather continuous. However, the basin of the Lower Balsas and thence to Colima will in all probability yield much of interest to whoever will brave these inhospitable and positively unknown regions".

When I undertook this journey in 1908 I was not disappointed. We may now assume the probable number of species for Michoacan to amount to 110, of which total are now actually recorded 86 species = 78 per cent. This is a rather satisfactory state of things, and I consider myself lucky in having found: 2 Urodela or newts, 16 Anura or frogs and toads, 3 tortoises, 1 crocodile, 30 lizards, 26 snakes—78 species, or 88 per cent. of the total now on record; and of these 78 species more than half, namely 43 species = 55 per cent., were previously not known to occur in Michoacan. The following were not expected to occur in this State:

Typhlops braminus *S. gadoviae*
Sceloporus cupreus *Ctenosura quinquecarinata*
S. melanorhinus *Phyllodactylus* sp. nov.

Rhinophrynus mexicanus *Rana palmipes*
Hyla mocquardi *Eupemphyx* sp. nov.
Leptodactylus caliginosus *Cinosternum cruentatum*

It was a pleasant surprise to find in the well-cared-for museum at Morelia a considerable number of species, with their proper localities, which had been collected by the expert and veteran herpetologist Dr Dugès, who lives in Guanajuato. This list was a welcome addition to the otherwise very scanty previous records from Michoacan.

The following 20 species have not yet been found in the State, but are most likely to exist there, to judge from their actual records in the neighbouring States of Guerrero and Talisco, and in the Territory of Colima.

Bufo simus *Chirotes canaliculatus*
B. intermedius *Glauconia dulcis*
Hylodes rhodopsis *Tropidonotus melanogaster*
Engystoma ustum *T. validus*
Rana montezumae *Zamenis mentovarius*
Eumeces furcirostris *Coluber chlorosoma*
E. brevirostris *Pseudoficimia frontalis*
Sceloporus acanthinurus *Leptodira albofusca*
S. variabilis *Manolepis putnami*
Laemanctus longipes *Ancistrodon bilineatum*

The total number of species known to occur in the Mexican Republic, from the north to the Isthmus of Tehuantepec (Yucatan, Tabasco and Chiapas are excluded from this census), is about 310. Michoacan therefore possesses 36 per cent., not a bad representation, considering how very much richer is the fauna of the south-eastern States; and it is worth noting, although it may be nothing but a coincidence, that the percentages of the Amphibia, lizards and snakes are 32, 37 and 33 respectively. These figures are as near one-third as can be expected, considering the unavoidable vagueness of such a census. The census stands as follows:

	Total number in Mexico	Computed number in Michoacan	Recorded for Michoacan		
Urodela	82	2	2	=	100 %
Anura		24	18	=	75
Lizards	100	37	30	=	84
Snakes	113	41	31	=	76
Crocodiles	3	2	1	=	50
Tortoises	10	4	3	=	75
Total	308	110	85		

Omitting species of insufficiently known distribution, the fauna of Michoacan is composed as shown below. The categories depend of course upon the principle which we select arbitrarily; the results alone attest to the value of the tentative selection.

Supposing we divide the southern third of Mexico into an eastern and a western half by an imaginary line which runs through the middle of the plateau from Zacatecas, by Guanajuato, to Mexico City and thence to the Isthmus of Tehuantepec. This line will, in the south-east, coincide with the somewhat intricate system of mountains which separates the Tierra caliente of the Atlantic States (Vera Cruz and part of Oaxaca) from that of the Pacific side, including the States of Oaxaca, Guerrero, Michoacan and Talisco.

There are then in Michoacan:

I. *Western Species*: about 50. Of these 40 are strictly Western, between the western and southern slopes of the plateau and the Pacific seaboard; the remaining 10 are essentially Western, but they extend on to the eastern half of the plateau and reach even into the State of Vera Cruz, by taking advantage of the mountains north of

Orizaba to Jalapa; but these species do not get into the Atlantic Hot Lands.

II. *Species which occur in the Atlantic and in the Pacific half*: about 38.

(*a*) Sixteen of these range right across the Republic, mostly (about 11) on higher levels; the others with a very wide distribution. But nearly all of them are recognisable as originally either of Sonoran or of Nearctic stock, or are Southerners which have ascended and become modified for high-level life.

(*b*) Twenty are essentially tropical species which have extended northwards from the Isthmus equally on the Pacific and on the Atlantic side.

It is obvious that the category II (*a*) is in some respects intermediate and is moreover a compound lot, therefore unsatisfactory. Adopting another principle we get:

A. Forty-four Pacific species, of which 38 are entirely Western, whilst 6 extend their range into the east. These Pacifics are composed of Southerners and Sonorans in nearly equal numbers, with a sprinkle of Nearctics.

B. Thirty-two species which occur both in the west and in the east, eminently within the tropical level. They are composed of 18 Southerners, 6 Sonorans and 8 Northerners. This shows, as was to be expected, the Southerners to be in the majority over the others, which are about equal. Since most of the Sonoran contingent are rather tropical it follows that these species have got to the east around by the south, say across southern Oaxaca into the State of Vera Cruz, whilst the Northern species simply had to descend from the plateau into the south and east.

C. Fourteen species which are dwellers on the plateau, or on the higher mountains which range across the country from west to east, south of the great plateau. These species comprise 8 Sonorans and 6 Nearctics. Perhaps it is better to combine these Sonorans and Nearctics with those of category B, in which case we get totals of $6 + 8 = 14$ Sonorans and $8 + 6 = 14$ Northerners, equally balanced.

The above calculations may be expressed as follows:

A. The Pacific lot (44), composed of Southerners and Sonorans in equal numbers, with a few of Northern origin.

B. The Ambilateral lot (46), likewise composed of the groups according to origin, but in very different proportions.

(i) Twenty-eight Sonorans and Northerners in equal numbers. Some of either are dwellers on the plateau or on the higher mountains, and they range across the country; e.g. *Sceloporus microlepidotus* and *Crotalus triseriatus*. Others are on the plateau but have extended their range thence into lower levels, even into the Hot Countries.

(ii) Eighteen Southern immigrants which have extended into the Atlantic and into the Pacific side.

Arranging the whole Michoacan fauna according to its provenance into Southerners (not necessarily the same as tropical species), Sonorans and Nearctics or Northerners, there are:

```
Southerners, total 39 of which 20 west, 19 also east
Sonorans,      ,,   31    ,,    18  ,,  13    ,,
Nearctics,     ,,   14    ,,     7  ,,   7    ,,
       Total        84          45      39
```

The preponderance of species which are restricted to the Pacific side is due almost entirely to the Sonorans. A similar census taken of Oaxaca and of Colima-Talisco shows that the Sonoran element decreases from the west towards the Isthmus, whilst the Southern contingent, very preponderant on the Isthmus, decreases towards the west until it is nearly exhausted about Mazatlan, where the fauna is composed almost entirely of Sonorans and Nearctics.

It is cheap to say that all this is exactly as it should be; that it could have been predicted from the configuration and other physical features of Mexico; that as a matter of fact the usual statement about the fauna and flora of that country is that it is the meeting-ground of the North and South American elements; and that the latter naturally extend northwards to the right and left of the great central plateau which separates the Pacific from the Atlantic States like a triangular wedge. Nobody thought of the very important existence of a third element, the Sonoran, which is endemic and archaic. This might also have been taken for granted in such a large country, but its existence was discovered only through much proving of statistics and laborious attempts to interpret the results. The value of such enquiries consists in the proof that the distribution of a fauna is not a haphazard phenomenon but intimately connected with, governed by, and therefore to be explained only by the physical features of the country, part of which features is its history. Thus we gain confidence in our method and are justified in applying its deductions to the elucidation of the faunistic problems in other countries.

The Atlantic tropical countries of Mexico owe their much richer fauna to a great extent to the number of Southern immigrants which are restricted to this side. We know, further, that Michoacan possesses at least 20 Southern species which are restricted to the Pacific side, and about as many Southern species which occur on both sides. It would be a cheap explanation to say that this is an accident without any deeper meaning, that the whole lot of Southern immigrants cannot be expected to have distributed itself equally. Is there any feature in which these Pacific Southerners differ from those which occur on both sides?

A review shows that of the Pacific Southerners, 14 are apparently restricted to the north of the Isthmus, or thereabouts, thence spreading westwards; while there are only 5 with a range which can be traced into Central America, *Ctenosura quinquecarinata* (to Honduras), *Mabuia*, *Himantodes gemmistratus* (to Costa Rica), *Phyllodactylus tuberculosus*, and *Hylodes palmatus* (to Nicaragua).

On the other hand, of the species which occur on both sides nearly all have a long Central American range, and reach even beyond into South America, while only a few (perhaps only *Rhadinea vittata*, *Trimorphodon upsilor* and *Leptodactylus albilabris*) are restricted to north of the Isthmus. In short, most of the Pacific Southerners are short-ranged, whilst nearly all 'ambilaterals' are long-ranged.

If, as we have reason to assume, these Southerners came from South America, or at least from Central America, into Mexico, they should all be long-ranged, because it would be preposterous to suggest that, whilst extending northwards, they gave up their old Southern homes, as if they migrated like a herd! The only reasonable explanation is that many of these which got beyond the Isthmus, changed there specifically—especially the offspring of those which found their way into the Pacific countries—in conformity with the difference of climate, which is much drier and with a prolonged period of drought. A scrutiny of the lists shows the differences in the fauna to be in most cases specific only.

It would be a bold attempt to suggest why some of the Pacific immigrants have remained unchanged. The tree snake *Himantodes gemmistratus* ranges from Costa Rica to the mouth of the Balsas, possibly because such an eminently arboreal snake can find the same congenial conditions wherever tropical shrubs occur; but why this species does not occur in the Atlantic States, although it is there represented by others, which likewise have a long range into Central America, is a question which would require more data and knowledge of their oecology than we possess. Of course accidents will happen, and this may be one, but it is not scientific to resort to this way gratuitously.

There are some remarkable cases of discontinuous distribution:

Rhinophrynus mexicanus, the sole species of its genus. Whilst this very sluggish termite-eating toad is common enough in the sweltering hot country of the State of Vera Cruz, up to an elevation of 1500 feet, it was unknown on the west side of the Isthmus until I found it in great numbers near the mouth of the Balsas River, in and near freshwater pools, where it attracted attention by its loud peculiar voice during the pairing season in the month of July. Of course it is most precarious to consider that a species is absent where it has not yet been found, but in the month of July, 1904, we camped at the same kind of place near the coast of Guerrero without coming across this toad, or hearing of it. Various naturalists have visited the lagoons near Acalpulco and many more have collected at Manzanillo, the port of Colima, but all in vain. Moreover, this is not an isolated case.

There are two species of the peculiar genus *Laemanctus*, iguanids with a curious helmet-shaped head; both are excessively rare, perhaps because they live up in the trees, where it is difficult to discover them. *L. serrapis* is known from the States of Campeche, Vera Cruz and eastern Oaxaca. The other species, *L. longipes*, was known from Jalapa only, that well-explored hunting-ground near Vera Cruz, often visited because it is the health station of that big insalubrious port. All the more remarkable is the solitary specimen which I found amongst a collection sent to the Field Columbian Museum at Chicago from the neighbourhood of Colima. Error, or mystification as to the locality, is quite excluded.

Ctenosura 5-carinata is another iguanid, hitherto known to range from Honduras into the hot parts of southern Oaxaca. It is arboreal but found also often on the ground; not at all shy, easily observed on the ground, or sitting upon a fence, or basking upon a low hollow branch. I found it plentiful inland of Salina Cruz, near the Pacific slope of Guerrero, but in great numbers almost everywhere in the depression of Michoacan, between Oropeo and Apatzingan. It is a species sure to be seen, easily caught and not likely to be ignored by the professional collector, since it is still rare in zoological museums.

Iguana rhinolophus, the largest lizard in Mexico, arboreal and aquatic, is known in the whole of Central America; in Mexico it is well known in the Atlantic States and in southern Oaxaca. Then comes a break, it never having been recorded from Guerrero, but recurring from the mouth of the Balsas, through Colima to San Blas; we can now add that it has ascended the Balsas valley and thence by its tributaries to the Jorullo. Most likely it ascends further up the Balsas valley and thus on to the northern or inland slope of the Sierra Madre of Guerrero.

The caiman, *Caiman sclerops*. Well known from South America through Central America into the State of Vera Cruz. As discussed elsewhere, its occurrence in the Balsas rests only upon circumstantial evidence, but can scarcely be doubted.

There are probably other cases of similar discontinuous distribution, occurrence in the Atlantic States and then within the basin of the Balsas with the intervention of the Pacific slopes of western Oaxaca and the bulk of Guerrero—for instance the pythonine *Loxocemus bicolor*, hitherto known from Guatemala and the Isthmus, and then from the mouth of the Balsas and from Colima. But this snake is small, scarcely two feet in length, and it leads a burrowing life, so that it is found by lucky accident only.

Each of the above cases, taken by itself, would not be of much weight, but taken together they seem to strengthen the evidence that for some species Guerrero and Oaxaca represent a gap of about 400 miles in an east to west direction. The absence of large enough rivers would readily account for the absence of the caiman, whilst the crocodile is not averse to brackish lagoons and even ventures into the sea. The hot and partly sand-covered depression in Michoacan repeats on a larger scale the partly sandy and equally hot conditions of the Tehuantepec river-basin, where the *Ctenosura* occurs, but precisely the same conditions, upon a small scale, exist on the coast strip of Guerrero. What may be the explanation of this discontinuous distribution we do not know, but it points with certainty to the conclusion that the respective species have been resident a very long time in the country, speaking geologically, dating back to a time when the whole of Central America was a much broader belt and the present Mexico extended considerably further west and south into what is now covered by the Pacific Ocean. There is not much doubt that this now sunken strip included, during the Miocene epoch, not only such islands as the Tres Marias, and the present Gulf of California, but also such far-off and so-called oceanic islands as the Revilla Gigedos and the Galapagos.

Appendix

EXTRACTS FROM THE LITERATURE CONCERNING JORULLO

Observations, Publications, etc., on Jorullo, arranged according to date

The date is the date of the visit, when this is known; otherwise it is the date of publication. The extracts which follow are arranged in the order of this list.

1759. Sáyago, Reports of October 8 and November 13
1759. Anzagorri, Letter of October 19
1766. Bustamante (see Clavigero)
1780. Clavigero
1782. Landivar
1789. Riaño
1789. Fischer
1789. Alcedo
1803. Humboldt
1804. Sonneschmid
1827. Burkart
1846. Schleiden
1853. Pieschel
1854. Orozco y Berra
1859. de Saussure
1870. Dominguez
1883. Leclercq
1888. Felix and Lenk
1906. Villafaña
1906. Ordoñez
1906. Hobson
1906. Cadell
1908. Gadow

Note. The two reports, of October 8th and November 13th, 1759, with which Dr Gadow begins this series of extracts, were first published by Manuel Orozco y Berra in the *Diccionario de Historia y Geografia*, tomo IV, año de 1854, p. 453, under the article 'Jorullo (Volcan de)'. They were found in a dossier in the Archivo General y Publico de la Nación. The first report, as published by Orozco, is unsigned and refers to the Administrator in the third person. The second is signed by the Administrator himself and from its opening sentences it is clear that if he did not actually write the first he was at least acquainted with its contents. It is a direct continuation of the first.

LITERATURE CONCERNING JORULLO   77

Concerning this dossier Dr Gadow has a few notes on a separate sheet:

'Orozco y Berra consulted it in 1854.
'J. de D. Dominguez consulted it in 1870. The *Boletin de Geografia y Estadistica, Mexico*, New Series, vol. II, 1870, pp. 561–565, contains the report as copied by Dominguez, Dominguez certifying the correctness of the copy.
'Ordoñez asked for it in vain in 1906.
'Reginald J. Tower asked for it in vain in 1910.'

P. L.

SÁYAGO. Report of October 8th, 1759, transmitted on October 13th by the Governor of Michoacan to the Viceroy of New Spain.

In the district of Ario D. José Andres de Pimentel, a resident at Patzcuaro, possessed a fine property, called the Hacienda de Jorullo and Presentacion respectively, for the production of sugar and cattle. There and in the neighbourhood, towards the end of June of the present year of /59 were heard and felt repeated subterranean rumblings or knocks, but without tremors, which caused the people who lived there much fear, and as these uncanny noises increased and were accompanied by slight shocks, the workmen forsook the hacienda to live higher up on the neighbouring hills.

It is an unalterable fact that this flight did arise not so much from the horrible increases of those noises and tremors as from a vague and spreading rumour that on St Michael's day Jorullo would come to an end.

On the 17th September, at 9 a.m., there was heard a formidable noise on the very spot of the Jorullo hacienda. This was repeated every moment and sounded like cannon discharged in the centre of the earth, so that the frightened people made for the chapel, to pray, but they had to fly to the hills because the incessant earthquake burst the chapel, threw down the shingles from the roof and did other damage.

Therefore the administrator of the farm sent to Patzcuaro for the Jesuit Padre Isidoro Molina, to celebrate Mass, etc., in order to appease the divine ire. P. Molina arrived on the 20th and on the 21st he began with a nine-days' Mass and confessional, and all this time the tremors and noises never stopped, until the 27th for a little.

There arrived the 29th of September, the much-feared day of St Michael, and at 3 o'clock in the morning, a quarter of a league to the south-east from the farm in the ravine called Cuitinga, broke out very dark and dense steam which rose up into the sky, having been preceded by three or four very sharp tremors, and soon after the appearance of the cloud a tempestuous and horrible noise was heard and flames of fire burst out high mixed with the cloud which every moment became thicker and darker. When the Padre Molina, the administrator and the other people saw this, they all, terrified, congregated in the chapel and, whilst they heard Mass, there began to rain water mixed with earth, to such an extent that, when the people came out, the ground was covered with much mud, and the roofs of the houses were much covered with the same; the sky was strangely dark and brown and to the thundering reverberations was added a strong smell of sulphur.

When the administrator saw this he mounted a horse and accompanied by the mayor-domo and some others, went to see the volcano; the administrator was the one who approached nearest to it, but he did not go the fourth part of the distance which there is from the farm to the volcano, because he had to run back on account of the frantic behaviour of the horses, moreover the road was already wiped out and what with the mud, the increasing vapour, the stench of sulphur and the darkness and noise the farm had to be abandoned.

On this 29th day of September fell so much water, sand and mud, that all the buildings were laid low and the farm was entirely spoiled, the damage amounting to more than 150,000 pesos, but the greatest misery was the pitiful plight of the hungry and shelterless farm-hands who had lost everything, and to see the cattle, mules and horses wandering about the hills without a twig to eat, or drowning without possibility of rescue from the inundated and sand-covered plain....

During the 29th and 30th the volcano threw out masses of sand, fire and thunder, without one minute's cessation. On the 1st of October burst forth a current of muddy water from the foot of a hill which lies on the other side of the volcano, on its southern side, and this current was so voluminous that it prevented one crossing over to the road which until this time would be used. On the same day was erupted a new outburst of sand, so hot that it set on fire whatever it fell upon; this flood of sand did not rise high, but just came to the surface and flowed downwards, following the current of the Cuitinga brook, which ran to the west. This was blocked completely; the sand, or rather the hot ashes, having run the distance of one quarter of a league, and there opened one after another, three mouths, not of fire but of vapour, throwing up high sods of dirt.

On October 2nd these features increased much, especially the outburst from the volcano of fire, which extended on the 3rd as a rain of sand as far as the Presentacion farm, two leagues [5 miles] to the west of the Jorullo farm, which between this day and the next, the

CONCERNING JORULLO

4th, was to a great extent inundated with soil, and it was lost entirely, the said sand having covered completely its sugar-cane fields, as the consequence of a furious earthquake which happened in the night of the 2nd.

On the 5th and 6th of October the annihilation of the Presentacion was completed and the natives of the village of Guacana, situated about half a league to the west from the Presentacion farm, fled into the neighbouring hills, their priest being the last to run away; and to-day this priest, with his Indians and the holy images rescued by them from his church is staying at the pass called Tamacuaro.

This sudden flight was caused not so much by the continuous rain of water and sand, but by a horrible spate of the stream which comes from the Jorullo and passes between Guacana and the Presentacion. The spate was produced not only by the rain from the sky, but by springs which opened from all the hills around. Now this river is so full that it is not only fearful to look at, but having levelled its old and deep bed, its waters are overflowing now here now there, causing much destruction to the fields of sugar cane and maize.

It is to be feared, if the fury of the volcano increases, considering what harm it has done already in such a short time, that all the valleys of the Jorullo, the Presentacion and the village of Guacana, may be turned into one big lake, what indeed they are almost already. There is first the incessant rain of sand which falls and gets mixed with the water, in addition to that which has levelled down, or filled up, the gorges and brooks, secondly the abundance of streams which all the neighbouring hills send forth, now swelling suddenly into full currents and then again suddenly running dry.

All these commotions have been witnessed and investigated by the P. Molina, the administrator, the mayor-domos and all the people who had come down to rescue as much as possible of the implements, furniture and stores of the hacienda, a very difficult and dangerous task amidst the everlasting earthquakes, storms and the darkness which continued ever since St Michael's day, and their fury may be inferred from the fact that the ashes from the volcano fell all around for more than 20 leagues.

On the 8th October something new happened: the volcano threw up a great lot of stones which fell down as far as half a league from its mouth, and which, as we found later on, were very soft and as if over-baked, or glassy.

Until this day, the 8th of October, the houses of the hacienda and the chapel are still standing, because they are quite new, built upon very strong foundations with buttresses of cut stone so that they could until now withstand the impact of the rain of ashes, mud and stones, although they are clearly cracked and are partly immersed in horribly stinking water.

The above report ends with the 8th of October; if there should happen any further news they will likewise be sent to the Governor of Michoacan. The messenger also takes a sketch or plan of the present condition and appearance of the volcano, etc.

Sr Orozco adds the following remark:

'The promised sketch was really sent and is still in existence, painted black and red with little care and scarcely serving to form an approximate idea of its object. There are also the instructions for the authorities to send in any further notes about the volcano, and here follows another diary of what happened':

SÁYAGO. Report of November 13th, 1759.

D. Manuel Román Sáyago, administrator-in-chief of the Haciendas Jorullo, Presentacion, San Pedro and dependencies, situated in the district of Ario, in obedience to the order received from the Governor of Michoacan...reports as follows:

...'Considering as sufficient the first report, which you have already sent to his Excellency the Vice-Roy, I now describe what happened after the 8th of October as follows:

On Tuesday, October 9, from 4 p.m. until the early morning of the following Wednesday alternated great noises and six sharp shocks and on the morning of the 10th the sky for a distance of three leagues was very obscure and there fell rain and sand especially towards the north-west carrying general destruction of the oak and pine forests, breaking all their branches and throwing many trees down to the ground. On this same day the sand fell as far as the Hacienda Santa Efigenia...about 4 leagues from the Jorullo, right in the direction of the wind. Since this day fell moreover great masses of rocks from the cloud, some of them as large as the body of an ox, which after having been shot up like a bullet, fell around the mouth of the volcano, and the smaller pieces, thrown up higher, came down at a longer distance and in such numbers that, scattering in the cloud, they looked in the day-time like a flock of crows and in the night like a crowd of stars.

Thursday, 11th, caused the same destruction in the hills as far as the sugar plantations of Nombre de Dios and Puruarán, both four leagues from Jorullo, but to the east. The globular cloud had namely changed its direction and spent its force chiefly upon the Cucha, a range of hills between Jorullo and Puruarán. And all along from the mouth of the volcano to this hill range raged a regular battle of flashes, fire and bombs, so that even at the copper mines of Inguaran

CONCERNING JORULLO

four leagues to the south of the volcano, the people were with difficulty prevented from deserting that place.

Friday, 12th, at 1 p.m. at a distance of 600 yards from the main crater broke out a new mouth, extending over the whole gorge downwards to the west, and this threw into the air a new and thick cloud of steam and such a great mass of hot water that it flowed for two hours like a spate, whereupon the gap closed and the water ceased.

Saturday, and Sunday 14th, the general darkness continued to such an extent, that I was not able to go to the hacienda, where I had intended with a number of men to rescue the holy image of Our Lady of Guadalupe, which as tutelary Saint and Patron stood in the Chapel. With this we succeeded not until Monday, 15th, when the cloud had shifted to the east and away from the farm. The image was intact, thanks to its curtain, but the other pictures, etc., were destroyed by the water and the ashes.... We also took down the bells from the tower; 150 people, men and women, helped to carry the sacred treasures to Cuarallo.

Tuesday, 16th, the volcano began to throw up sand or ashes already dry and apparently coarser; the fire was fiercer and the springs of water had run dry; the sky of the colour of straw, still discharging ashes accompanied by noise, without interruption until Saturday, 20th; during the whole of the following week, until Saturday 27th, nothing new happened except that now and then a wind arose and spread the dry ashes, which the volcano brought forth, over the whole cattle farm of San Pedro, about four leagues to the south west of the Jorullo, even to Oropeo, eight leagues farther west and to the farm of Guadalupe farther still in the same direction. The cattle could find nothing to eat, the trees and shrubs being destroyed and the leaves covered with ashes, and nothing to drink because during all this time the water was rendered unfit by mud and sulphurous matter. The same happened at Zicuiran, Cunguripo and Guatziran, all situated towards the west at distances of 10 and 12 leagues.

At the end of this week the priest of Guacana came down with all his Indians in order to take the holy vessels, etc., out of his church, the roof of which had fallen in, and to convey them to Churumuco, some 15 leagues to the south. I have to report, first, that until Saturday, 27th, the volcano did not pause one single minute spitting out its ignited matter and ashes with tempestuous noise, but it yielded no more water...; secondly that the poor Indians did splendid service during the transport, although without food, and suffering much from their eyes which became inflamed by the dust.

Sunday 28th, at daybreak, the cloud of the volcano was slender and ashy coloured; when it was lighted up by the sun it became white like a cotton-pod, and the noise from its mouth had changed too; every now and then it thundered like a cannon, discharging a great lot of stones without ashes; in between these eruptions it sounded like the bellows of a smithy and then again like a mortar.

And the flames shot up to such a height that they lighted up the mountains a dozen leagues around. In this manner things went on until Thursday, November 1st, during which days we had at least the relief of being able to see the sun. But by the 2nd the sky had thickened again and the cloud had enlarged and returned to its old condition. No improvement took place until Wednesday, Nov. 7th, when D. José de Arriaga the Chaplain, and the Administrator of the Hacienda Nombre de Dios arrived on the scene in order to exorcise the volcano. This did not come off there, because a sudden fierce rumble from the crater made them run to a hill half a league further, whence the monk applied the exorcism to the volcano. But on Thursday the 8th it was worse, and on Friday it smothered the whole valley of Urecho, 10 leagues to the north-west. The darkness was worse than ever, accompanied by furious earthquakes to which were added for the next four days some hurricane with thunder, lightning and downpours of rain all over the neighbourhood.

To-day, Tuesday 13th, I went down (from the hills) for a new ocular inspection of Jorullo to find out whether that pitch or lava has run, about which His Excellency has asked in particular; what I have found is this, that all the old brooks of the hacienda, which, as I have mentioned before, now run on the top of the sandy plain, are quite thin and clear, in parts quite like crystals, as thick as a finger; they being so clear, one can see at their base a white kind of putty, something like dissolved lime, with a tint of yellow, and as thin as a sheet of blotting-paper; and at the margin this stuff is transparent with the look of mother-of-pearl and fat; if one tried to take it up with the fingers, it falls to pieces or dissolves at once in the water. Therefore I could collect none of this stuff, only the grains of saltpetre which seemed to me more infected by it, if taken from the scum. Of these I am sending about one pound to you, Sr Alcalde Mayor, which may be examined at your pleasure. I have no doubt that if these ashes be mixed with water and are allowed to settle, the said 'beton', pitch, or putty will come out. Please let His Excellency know in answer to his special question, that I neither have any knowledge of so-called lava, nor have I anyone to tell me what stuff it may be; but whatever it may be here does not run, or flow, anything else.

Here ends the diary of what has hitherto been seen and observed of the never 'bien ponderado volcán', and the damage it has done, until this Tuesday, the 13th of November, and in conclusion I shall only add a few remarks.

First: It did not break out on the top of some hill, as was the case with the other volcanoes which one sees in this kingdom, but it broke out in the deepest and level part of the narrow valley Cuitinga, which stands at the foot of the high hill of Cucha.

Second: The difference in the sounds which it produced since the day on which it burst, and especially when its cloud had become large and produced several storms of lightning, sparks and fiery explosions.

Third: Having belched forth such a countless mass of red hot stones so that around its mouth was formed a circular wall or ledge, which is already higher than 300 varas, and surpasses the others which stand on the sides of the valley, which latter it has filled up and disfigured.

Fourth: That with all this stress and ruin not one of the many unfortunate people has lost his life....

All this, with the related circumstances, I, the said administrator, have seen, observed, and studied, so that I can affirm this diary to be true and nothing but the truth, as I am and have been the most immediate eye-witness of all that has been mentioned in this report, which I sign here at the Cuaralla farm, on the 13th day of November 1759. *Manuel Román Sáyago*."

JOAQUIN DE ANZAGORRI, *curé* of Guacana, wrote a letter to his Bishop in Morelia, dated Guacana, October 19th, 1759.

This letter was discovered in the episcopal Archives in the year 1830; and was used by Humboldt in his *Kosmos*. Extracts from it have been published by Burkart in his *Aufenthalt und Reisen in Mexico*, Bd. I, S. 230, 1836; and by Orozco in *Bol. Soc. Geogr. y Est.*, t. v, 1857.

The only passage worth mentioning is that the eruption had ruined the farm to such an extent that the buildings, sugar plantations and trees were overwhelmed by the masses of sand, ashes and water which the volcano vomits. 'So much ash is still falling that it has completely covered the fields.' The Guacana river is much swollen.... Anzagorri was an eye-witness to a certain extent, from a distance. The report of October 8th tells us that the priest ran away on October 6th to the Tamacuaro pass and stayed there; on about the 27th he returned to

Guacana to take his images to Churumuco. It is therefore somewhat difficult to understand why he dated his letter Guacana, October 19th.

F. S. CLAVIGERO. *Storia Antica del Messico*. T. I. Cesena, 1780.

The Abbé Clavigero, a native of Mexico, where he lived for 40 years before he took exile in Italy, has written a most interesting work on the people and the natural history of his country. This work has been translated into Spanish by J. de Mora, published in London 1826; an English translation by Charles Cullen is dated London 1787.

The following short account of the Jorullo is relegated to a footnote on p. 13 of the Spanish edition, p. 42 of the Italian edition.

The Juruyo, situated in the valley of Ureco in the kingdom of Michoacan, was before the year 1760 nothing but a small hill with a sugar mill on the top. But on the 29th September of that year it burst with furious earthquakes, which ruined the sugar mill and the neighbouring village of Guacana; and since that time it has not ceased emitting fire and burning stones, *from which three high hills have formed, the circumference of which was even then of about six miles, as testified by a report which in 1766 Don Juan Manuel de Bustamante, Governor of that province and a reliable witness, made to me.* The ashes thrown out by the outburst reached Queretaro, a town 150 miles distant from the Juruyo: an incredible thing but well known in that town, an inhabitant of which has shown me the ashes which he had collected in a piece of paper in the town of Morelia, 60 miles distant; the rain was so abundant that the courtyards had to be swept two or three times a day.

Note. The original Italian work is not accessible in Cambridge and Dr Gadow translates from the Spanish edition of J. de Mora. This version is open to the interpretation that the Governor did not speak from personal knowledge but merely certified the official nature of the report which he communicated to Clavigero. Dr Gadow therefore took some opportunity of consulting the original work and made the following extract, which he evidently intended to incorporate:

...dai quali si son formati tre monti elevati, la cui circonferenzia era già di sei miglia incirca, atteso il ragguaglio che nel 1766 mi fece il cavaliere Giovanni Emanuele di Bustamante, governatore di quella provincia, e testimonio oculato.

Mr Edward Bullough has kindly furnished a translation, which has been substituted above, in italics, for the corresponding part of Dr Gadow's translation from the Spanish. Mr Bullough adds that even in this there is some ambiguity. 'Testimonio oculato' in modern Italian would mean a 'reliable witness,' in eighteenth-century Italian it might mean an 'eyewitness.'—P. L.

RAFAEL DE LANDIVAR. *Rusticatio Mexicana.* Bologna.... Ed. altera, 1782. Liber secundus: *Xorulus.*

This poem of about 350 Latin hexameters is the first printed account of the eruption of the Jorullo volcano. Most of the space is devoted to long-winded accounts of the prophecy and the terrible scenes caused by the catastrophe, of which itself the following are the important passages:

> Cum subito tellus horrendo rupta fragore
> Evomit Aetnaeas furibunde ad sidera flammas,
> Ingentesque globos cinerum, piceasque favillas,
> Obscura densans totum caligine coelum.
> Flammea saxa volant rutilis decocta caminis
> Et crebro tellus casu tremefacta dehiscit.
>
> Tot vero interea flammatae fragmina rupis
> Impatiens ructat monstris faecunda vorago
> Ut saxum saxis, ac rupes rupibus addens
> Ingentem mediis montem[1] glomeraverit agris.
> Una tamen cunctis cum non satis esset abyssus
> Quatuor hanc circum, sectis compagibus, ora
> Ardenti Vulcanus edax torrente recludit....
> Sic laetos quamquam spoliavit germine campos,
> Terraque per lustrum nullis fuit apta serendis
> Fructibus; at vero ex illo tot tempore factus,
> Antiquum ut vincat praesentia commoda damnum.

This reasonable and matter-of-fact description is difficult to reconcile with Humboldt's remark (*Kosmos*, Bd. IV, S. 563):

[1] Footnote by the author: 'Congesta saxe montem in medio vallis afformant altitudines ad milliaria tria'.

'Landivar, the poet, like Ovidius enthusiastically devoted to our theory of elevation, describes the colossus as arriving to the full height of 3 milliaria....' There is not a single line in the whole poem which speaks of a rising of the ground!

RIAÑO. 'Superficial y nada facultativa descripción del estado en que se hallaba el volcan de Jorullo la mañana del día de 10 de Marzo de 1789.' *Gazeta de Mexico*, 5 de Mayo 1789, t. III, num. 30, pp. 293–7.

Under the humble title 'superficial and not at all professional account of the state of the Jorullo volcano on the morning of March 10, 1789', has been published the account of the inspection made by Lieutenant-Colonel Antonio de Riaño, Governor of Michoacan, accompanied by the German mining expert Franz Fischer, and Ramón Espelde, a gentleman who had previously ascended the volcano, the first ascent ever made. J. M. Marroquin and S. Schroeder, a German working man, were also of the party.

...The mouth of the volcano is on the top; the crater is like an inverted cone at the bottom of which is a kind of gorge, apparently 900 yards long from the north to south and little more than 650 from east to west. In the latter direction the hole is shallower and less steep owing to the walls having partly fallen in. The concave contours of the crater, from north to south as well from east to west are, at its lips, prolonged into four prominences which can be seen from below. The bottom of the gorge along the concave line from north to south which is the deepest, is of a colour somewhat between ox-blood and scarlet, whilst the shallower east to west portion is less vividly coloured.

For the last 20 yards from the top one has to walk over baked, coloured stones which look like clinkers of iron ore.

The whole hill of the volcano is bare and has only here and there some small trees, locally called Ortiga silvestre, resembling the new shoots or branches of the Fig-tree; but its roots are so shallow that they assist one only if he grasps the trunk close to the ashes, otherwise they tear out and loosen without affording the slightest help. The hill is further covered with some patches of so-called Zacate, but this grass has such weak and shallow roots that it is torn out at once.

Before this terrible hill and its companions burst into appearance, earthquakes and underground noises were noticed repeatedly, and on the day of the frightful event it was observed that the surface of the

CONCERNING JORULLO

ground rose perpendicularly, more or less bulging up and forming huge bladders, the largest of which is to-day the hill of the volcano. These swellings, big bladders or cones of various sizes and shapes, burst and threw out of their mouths boiling mud (boiling hot soil) and stones more or less caked (baked and molten). This fell at prodigious distances as can even now be ascertained clearly by the black ashes which cover the roads six leagues off. The strongest and most copious eruption went towards the west and north-west. It is said that the eruptions continued for the whole of the first 16 or 17 years, and even now they are sometimes repeated, although to a much smaller extent. At the present time the other bladders or small swellings in the neighbourhood of the volcano are smoking very little and many of them are quite extinct, even broken down and apparently for ever. The same will happen in time to the volcano itself, which is still alive and terrible, because the vertical walls of its crater are threatening to fall in, and when there is no more molten matter to come up, the whole hill will sink in and remain like other extinct volcanoes, ages ago a terror to the people.

For the distance of a league from the base of the volcano, and the swellings near it, one encounters the bulky and heavy erupted fragments which compose what the natives call Mal Pais. Two roads pass over it through loose sand more or less burnt, much like dark blackened ashes. The tracks lead up and down various mounds and across a terrain...well remembered by old people who had seen the mill and the sugar-fields where is now a horrid-looking waste of fine burnt sands.

More than half a league to the west of the volcano stands a small knoll, quite hollow and covered with hard baked soil; it sounded like a drum when we stood on it. This specimen of bladder or swelling differs most from the others whilst it rather resembles the volcano by its activity, since it emits thick smoke through several holes which, like quite a number of other chimney-like cracks, are distributed over its surface. On some spots the inner fire is strong enough to scorch the feet and one cannot hold his hand to the holes of these chimneys on account of the moist heat.

A few yards from the road, where this passes nearest the volcano close to its base, is a flat spot, covered with a crust of hard baked soil which forms the roof of a vault, at least it sounded like one, and hot steam came out of some cracks. There are various similarly behaving spots all over the three or more miles of the Malpais.

On the north-east base the big volcano is entered by a brook and when this carries much water, as in the rainy season, the internal fire increases and thicker steam is emitted.

A league and a quarter from the volcano, still on the Malpais, is a narrow defile with several almost boiling hot springs; farther down these are used for bathing by invalids, some of whom recover their health.

An indescribable feeling of fear and fright affects one near the volcano and anywhere on the Malpais, but as soon as he has arrived at the top, he is wrapt in admiration of the view which comprises the very foot of the mountain and the bottom of the crater.

F. FISCHER. *Schriften d. Gesellschaft d. Bergbaukunde.* Bd. II, Leipzig, 1790, S. 443. A shorter extract also in *Köhler's bergmännisches Journal*, IV. Jahrgang, 1. Bd. (1791), S. 325. Letter dated Guanjuato, April 15th, 1789.

Ungefähr 30 Meilen von Valladolid gegen Süden befindet sich ein Vulkan, den ich mit dem Gouverneur dieser Provinz, D. Antonio Riaño, einem Manne von vielem Verstande, der mich auf dieser Reise begleitete, bestieg. Dieser Vulkan ist vor 30 Jahren auf einer Fläche entstanden, auf welcher mehrere Zuckerplantagen angelegt waren. Man verspürte anfangs ein gewaltiges Erdbeben, welches die Einwohner dieser sonst so fruchtbaren Gegend veranlasste, die Flucht zu ergreifen; dann öffnete sich die Erde und warf so viel Steine und Asche aus, dass viele Meilen weit sich Niemand nähern konnte; *die Hauptverwüstung aber geschah in einem Umkreise von 1–1½ Meilen, den man nicht ohne Schauder betreten kann.* Die ersten 4 Jahre waren die fortwährenden Ausbrüche des Vulkans sehr heftig. Nachher geschahen sie noch 11 Jahre mit mehr oder weniger Heftigkeit. Jetzt raucht dieser Vulkan nur noch; zur Regenzeit bemerkt man Erdbeben und hie und da einige unbedeutende Erdbeben. Der ganze Vulkan hat die Figur eines abgestumpften Kegels. Seine Höhe beträgt an der Morgenseite, von der wir ihn bestiegen, 5–600 Schuhe mit einem Verflächen von 45 Graden. Von der Süd- und Abendseite ist er etwas höher. Wenn man hinaufkommt, passirt man eine Art von Fläche voll Spaltungen, die einen Schuh und öfters mehr weit sind, aus welchen Rauch und Dampf hervorsteigt. Diese Fläche macht rund umher den Kranz des Kraters aus, dessen Schlund ganz eingerollt und mit senkrechten oder überhängenden Steinwänden, die gelb und weiss beschlagen sind und ununterbrochen rauchen, umgeben ist. Die Weite des Kraters beträgt von Süden gegen Norden 800 und von Osten gegen Westen 400 Schuh. Man findet hier keine eigentlichen Laven, sondern halbgeschmolzene Steine, die mit verschiedenen Salzen zusammengebacken sind. Gegen Abend findet man noch an verschiedenen Orten brennende Stellen, und am Ende der Verwüstung, welche man das *Ueble Land* (Mal Pays) nennt, trifft man viele siedend-heisse Quellen an.

A. DE ALCEDO. Article 'Xurullo', in *Diccionario geografico-historico de las Indias*, Madrid, 1789, pte V, pp. 374–5. Reprinted by Villafaña, p. 99.

Alcedo was not an eye-witness; whether, or when, he visited the scene of action, is not known. Xorullo is a Tarascan word, meaning Paradise. The eruption transformed the once pretty valley completely.

All is blackened by the continuous fire, covered with shapeless rocks and ashes, the trees are burnt, the ground is full of cracks and holes, and a high mountain with a volcano stands now where there was once a plain. The stream which formerly fertilised the valley is now called the Salto, The Jump, because its water is so hot that riders have to jump for fear of being scalded, and this they have to do on the way to the copper mines of Inguaran.... When the earthquakes began here, the eruption of the Colima volcano stopped, and it seems that the matter which was blocked in the entrails of the earth found a new outlet in Jorullo, although the distance is so great.

HUMBOLDT. *Essai politique sur le royaume de la Nouvelle-Espagne.* Paris, 1811. *Essai géognostique sur le gisement des roches.* Paris, 1823. *Kosmos*, Bd. IV, Stuttgart, 1858.

Humboldt paid a visit of two days to Jorullo, on September 18th and 19th, 1803, just 44 years after the eruption. His historical information at that time rested upon Riaño's description in the *Gazeta de Mexico* of 1789, Fischer's letter of the same year, and upon what was told him by Espelde, a gentleman who at the time of Humboldt's visit lived at the Playa. Lastly, the natives were cross-questioned; some of them may have been eye-witnesses, but 44 years is a long time for illiterate Mexican Indians. For the account in the *Kosmos* he also had at his disposal the letter of Anzagorri.

The following passages are from the *Essai politique*, t. I, pp. 250 et seq.

...in the night of the 28th to the 29th a horrible subterranean tumult manifested itself again. An area of 3—4 sq. milles [2·3 sq. miles] which is called the Malpais, arose like a bladder. The original limits of this elevation may still be recognised by the broken strata at the edge...the convexity of the elevated area increases progressively towards the centre to a height of 160 metres. ...In the middle of the pushed-up area, on a crack running from NNE to SSW came six large hills, all elevated from four to five hundred metres above the ancient level of the plain,

the highest of these hills being the volcano of Jorullo.

In the *Essai géognostique*, p. 353, the lifted-up ground is described (at the abrupt edges of the Malpais) as composed of banks of black and brownish-yellow clay, covered on the top by but little volcanic ash. Further, it is stated that the convexity of the Malpais has not been produced by any heaping up of clinkers and volcanic ejections.

The following extracts are from *Kosmos*, Bd. IV, pp. 334 and 562:

The place now occupied by the great volcano was originally covered with shrubs of Guayava (*Psidium pyriferum*), much esteemed by the natives for their delicious fruit. Labourers of the Hacienda de San Pedro Jorullo had gone out to gather them, and when they returned to the farm they caused astonishment because their large straw hats were covered with volcanic ashes.

To judge from a letter written three weeks after the first day of the eruption by the Padre Joaquin de Ansagorri (discovered in the episcopal archives of Morelia in 1830), the Padre Isidoro Molina of the Jesuits' College in Patzcuaro was dispatched to give spiritual comfort to the people at the Playas de Jorullo, much frightened as they were by the subterranean noises and the earthquakes, and that he was the first to appreciate the increasing danger and thus brought about the saving of the whole little community.

After a few hours the black ashes were already lying a foot high; everybody fled towards the heights of Aguasarca, a little Indian village, 2260 feet above the old plain of the Jorullo. From these heights, so the tradition says, a larger stretch of the land was seen to be in a frightful outburst of fire, and—to use the words of those who witnessed the rising of the mountain (Berg-Aufsteigen) 'in the midst of the flames there appeared a large shapeless lump (bulto grande) like unto a black castle (castillo negro)'...

According to the tradition, spread far but unvaried among the natives, during the first days the eruptions of large blocks of rocks, cinders, sand and ashes were always accompanied by an outburst of muddy water. In the afore-mentioned report of Oct. 19, 1759—the writer of which describes these incidents with great local knowledge, it is stated especially 'que espele el dicho volcan arena, ceniza y agua' (that the volcano brought up sand, ashes and water). All the eyewitnesses narrate (I am translating from the description of the state of the volcano on March 10, 1789, by Colonel Riaño and the German Franz Fischer, who had entered the Spanish service as a Commissioner of Mines) that, before the terrible mountain appeared the shocks and the underground noises were more frequent, but that on the day of the outbreak, the flat ground was seen to rise vertically (antes de reventar y aparecerse este terrible cerro...se observo que el plan de la tierra se levantaba perpendicularmente) and the whole expanded more and more, so that bladders (vexigones) appeared, the largest of which is to-day the volcano (de los que el mayor es hoy el cerro del volcan). Later on these pushed-up bladders burst, some of them of rather conical shape, and of various sizes (estas ampollas, gruesas vegigas ó conos diferentemente regulares en sus figuras y tamaños, reventaron despues) and they expelled from their openings boiling hot mud and clinkers, which covered with black stony matter may still be found over enormous distances.

These historical reports, which indeed might well be more explicit, completely agree with what I myself heard by word of mouth from the natives, 14 years after Riaño's ascent. When I asked them, whether the mountain-castle had become higher in the course of months or years, or whether it looked from the first days like a high peak, I could get no answers; they further denied any eruptions during the first 16 or 17 years, a statement made by Riaño....

The plain of the Playa is 2400 ft above sea-level.

...The rounded convex portion of the lifted-up plain measures about 12,000 feet across, therefore with an area of more than one-third of a geographical square mile...From my standpoint at the Playa or on the Cerro del Mirador in the early morning the black volcano looked very picturesque, towering above the innumerable white columns of smoke emitted by the 'little ovens' (hornitos). The houses of the Playas as well as the basaltic hill Mirador are at the level of the old non-volcanic, or rather not lifted-up ground. Its pretty vegetation of masses of *Salvia*, blooming beneath the shade of a new kind of fan-palm (*Corypha pumos*), and a new kind of Alder (*Alnus jorullensis*) is in contrast with the barren, plant-less appearance of the Malpais. Comparison of the barometer readings at the point where the elevation of the Playa begins and at the foot of the volcano gives a vertical difference of 444 feet.... At a distance of 20 feet from some of the hornitos, where none of their steam could reach me, the temperature was still $42°·5$ and $46°·8$, whilst the true air-temperature of the Playa at the same time was scarcely $25°$....

In the middle of the lifted-up area, which is covered with hornitos, stands a relic of that old hill, against which had been built the houses of the farm San Pedro. The hill, indicated in my map, forms an east to west ridge and it is surprising that it should have been preserved at the foot of the big volcano. Only part of it is covered with dense sand (calcined rapilli). The cropping-up basaltic cliff, upon which are growing extremely old stems of *Ficus indica* and *Psidium* (uralte Staemme), is with certainty to be considered as pre-existent to the catastrophe, just like the Cerro del Mirador and the high mountains which surround the plain towards the east. ... The southern volcanic hills [Enmedio and del Sur]... are covered entirely with grey-white volcanic sand.

The volcano proper stands 667 toises (4002 feet) above the level of the sea, 180 toises (1080 feet) above the Malpais at the foot of the volcano, and 263 toises (1598 feet) above the old bottom of the Playas.

The ascent was made by Humboldt, Bonpland and C. Montufar on September 19th, 1803, right across the black mass of lava.

After we had ascended 667 feet, to the top of the lava-flow, we turned to the white cone of ashes, where owing to its steepness and our frequent slipping down, we ran danger of being hurt by the sharp edges of the lava....

SONNESCHMID, *Beschreibung der vorzüglichsten Bergwerksreviere von Mexico*, 1804, S. 325, says:

Die nachstehende Erzählung ist der Bericht einer sehr glaubwürdigen Person, die damals auf dem Landgute wohnte, das durch den vulkanischen Ausbruch sehr gelitten hat.

Da nun diese Begebenheiten beinahe einen Monat gedauert hatten, so wurden die Erdbeben häufiger und das dabei entstandene Getöse noch viel schreckhafter, so dass bei dem Anfange jedes Erdbebens während einer halben Stunde ein so entsetzlicher Lärm ausbrach, als wenn alle benachbarten Berge zusammen stürzten, *und zugleich hatte es den Anschein, als wenn der ganze Erdboden gehoben würde.* Auf solche Weise ging es fort, aber so heftig und so oft wiederholt, dass in jeder Minute 4, 6 und 8 Schläge gehört wurden, gerade als wenn sich zwei Kriegsschiffe kanonirten. *Endlich* (29 September, 3.30 a.m.) *zerplatzte der Vulkan,* in der Ecke von Cuitinga eine Viertelstunde weit von dem oben genannten Gute und dabei wurde der Berg von San Francisco mitten durchgespalten und auseinander getheilt.

Die Erdbeben mit dem unterirdischen Donner und Poltern hatten also 3 Monate und 5 Tage gedauert; sie wurden alle Tage heftiger und zuletzt war das Getöse ununterbrochen fortwährend bis zu dem völligen Ausbruche des Vulkans.

Sonneschmid was in Mexico in the year 1790, before and after. He did not visit the volcano. His introductory remark about the very credible person, who lived at that time at the hacienda, is badly expressed. He can refer only to Espelde, who was settled there in 1789 and is supposed to have made the first ascent in 1780, but never posed as an eye-witness of the eruption.

BURKART. (1) *Aufenthalt und Reisen in Mexico, in den Jahren 1825 bis 1834.* Stuttgart, 1836. (2) *Karsten's Archiv f. Mineralogie, usw.* Bd. v, S. 189. (3) 'Ueber die Erscheinungen bei dem Ausbruche des mexicanischen Feuerberges Jorullo im Jahre 1759', *Zeitschr. d. geol. Gesellsch.* Bd. IX, 1857, S. 274–99.

Burkart was a geologist, who travelled much in Mexico and died as Professor of Geology at Berne. In the month of January 1827 he traversed the whole district of the Jorullo, on his way up from the south coast to Patzcuaro. His account of the volcano begins on p. 224 of vol. I of his book.

In the neighbourhood of the Rancho Cayaco, still six leagues distant from the volcano, he met with the first deposits of ashes. 'The ground begins to be covered, sometimes several feet deep.' 'Northwards from this Rancho the ashes cover all the live rock, and only near the Rancho Joya de Aloarez, and a little further north, the grey basaltic rock crops up.' On p. 227 of his book (vol. I) he says that a great number of hornitos, observed by Humboldt, have disappeared—scarcely 24 years after his visit—owing to the very strong rains and to '*the vegetation which is spreading more and more every day*'.

Burkart being a supporter of Humboldt's theory of upheaval or elevation, wrote another article (*Zeitschr. d. geol. Gesellsch.*) in order to refute some of Schleiden's adverse criticisms. On p. 281 he describes the edge of the Malpais.

I have examined the wall-like margin for a long distance near the western side, and having worked my way through the vegetation which is already growing luxuriantly, I was able to see that it rises up almost vertically, like a sharply cut wall, 20 to 30 feet high, without step-like ledges, so that I scarcely anywhere could climb on to the Malpais itself. The wall consists of light grey not at all dense basaltic rock, with many granules of olivine, and it was divided into several banks by more or less undulating, almost horizontal cracks. Nowhere have I observed a fissure dividing it at the bottom from the underlying plain or step-like ledges, and above all, there was nowhere such a rough, torn, swollen up or crinkled face as would be presented by the under surface of a viscid flow of lava which has become solidified during its progress. After several futile attempts to climb on to the Malpais I succeeded at last at some lower places. I found many of Humboldt's little cones completely gone and the others much changed in shape.

Only a few of the cones still showed a temperature higher than that of the air and scarcely any exhaled watery vapours. The little cones near the margin of the Malpais consisted mostly of porous basaltic lava, those nearer the principal volcano of a brown-red conglomeration of roundish and sharper-edged fragments of stony basaltic lava, but loosely connected without any obvious binding matter. The conical shape, as sketched by Humboldt, had quite disappeared, whilst it was better retained by the basaltic cones. Only the queer-looking marks of long-drawn-out, concentric rings, 8 to 10 inches distant from each other indicated the former existence of cones in the vicinity of the main volcano.

I climbed to the crater over loose pieces of various kinds of lava... small craters still exhaling steam of 45 to 54 degrees which kept the ground warm.

The hot springs indicated only 38 degrees, 14·7 less than found by Humboldt, and the Malpais showed no longer a temperature higher than that of the air....

Burkart devotes the last dozen pages of his article to a refutation of Schleiden's natural, in reality quite obvious, explanation of the whole Malpais as the result of several successive and partly superimposed enormous flows of lava. He even goes so far as to say that such a flow of lava cannot have taken place as it is not mentioned in Anzagorri's letter!

E. SCHLEIDEN. *Fortschritte der Geographie und Naturgeschichte, von Froriep und Schomburgk*, Bd. II, 1847, Lief. 16. Two plates and several text-figures of the hornitos.

CONCERNING JORULLO

Schleiden visited the Jorullo on February 7th, 1846. Starting from Patzcuaro on February 4th, he experienced a thunderstorm, unusual at that time of year, also 'thunder' from the volcano and later he ascertained that those noises had really come from the crater. The Hacienda Tejamanil was surrounded by bananas, palms, pineapples, higueras (*Ficus*) and parrotillas with twisted pods (*Acacia*), bananas growing beneath the shade of long-leaved pines. There were also granadilas de China, a gigantic Passiflora with red-yellow, egg-shaped fruits, climbing into mighty oaks; at the brooks very old sabinos (*Cypressus disticha*). The sketch of the Jorullo, made at the palm-grove of the Tejamanil, he warrants to be quite correct.

Although I counted more than 100 fumaroles in the coolness of the morning, only two of these were on the Malpais, i.e. upon the enormous mass of lava spread over the SW portion of the valley. This main flow extended into the neighbourhood of the present bed of the brook which flows through the valley, and it ends as a steep, partly vertical wall, 20–30 ft. high, made up of blocks mixed with layers [Schalen]. Over the top of the Malpais these broken masses or blocks are most irregularly distributed and from the core of the hornitos which may be round, elongated like a back, or contorted.

Upon this first outflow of lava followed a rain of sand and ashes: uniformly covering the whole, rough surface with thinner layers of fine ashes from grey to dark red or black. In some parts the rain water has by now washed down masses of ashes and sand into the depressions of the terrain, filling them to a depth of 30 feet.... The greatest number of fumaroles were observed at the southern margin of the higher portion of the third mass of lava erupted towards the WNW of the volcano.

About 40 gas and steam vents existed on the rim of the crater. Their vapours dissolve the lava blocks into a pale yellow earthy or clay-like matter and cause the cliffs to be covered with yellow efflorescences, but there is no sulphur.

The upper part *of the volcano*, consisting of black lava, *and all the lava streams, excepting the first flow* [which forms the western half of the Malpais] *are still quite free from vegetation, but the sandy slope of the main cone and the Malpais show already considerable growth* [ziemlich bewachsen]. Some kind of *Mimosa*, not very high, and *Guayava* trees, remarkable for their fruit, are the most important; a pine[1] near the summit seems to me the most interesting of the members of this vegetation.

C. Pieschel. 'Die Vulkane von Mexico'. Fünfter Artikel (Jorullo) in *Zeitschrift für allgemeine Erdkunde*, Bd. VI, Berlin, 1856, S. 489.

Im Monat Juni 1759 liess sich unterirdisches Geräusch hören, welches von häufigen Erderschütterungen begleitet wurde. Im Anfange des September schien eine völlige Ruhe einzutreten, bis in der Nacht von 28 zum 29 dies unterirdische Geräusch auf eine erschreckende Weise sich erneuerte, *und der Boden auf einem Raume von 3–4 Quadratmeilen sich erhob, dessen höchster Punkt nach und nach auf 480 Fuss emporstieg*. Augenzeugen auf der Höhe von Aguasarco versichern, dass an einer Stelle eine halbe Meile im Geviert Flammen aufgestiegen seien, die glühende Steine und dicke Rauchwolken zu einer ungeheuren Höhe ausgeworfen, und dass die erweichte Erde wie ein bewegtes Meer sich erhoben habe. Die beiden Flüsse Cuitimba und San Pedro stürzten sich in die brennenden Schlünde und gaben den Flammen, die man in der Stadt Pazcuara, 20 Leguas weit und 1400 Meter über der Ebene des Jorullo, gesehen haben will, neue Nahrung. Diese gewaltigen Eruptionen haben bis zum Februar 1760 gedauert. In den darauf folgenden Jahren haben sie allmählig ganz nachgelassen.

Pieschel does not mention the source of the above account, but he atones for this by a vivid description of the physical features of the Jorullo, which he visited on January 25th, 1853. The following remarks refer to the state of the vegetation. The slopes of the Sierra bordering the district in the north-west were covered with fan-palms, oaks and conifers. The sand-covered slopes and plains towards the south and east were covered with the same small tree-like fan-palms. The plains, deeply covered with ashes, had been converted into fertile fields of water-melons and indigo, while sugar-cane was cultivated on the Playa. The Pedregal between Agua Blanca and the Playa, 'composed of sharp pieces of lava, was filled at many places with fine, black volcanic sand, and already covered with luxuriant vegetation'.

[1] 'Eine Fichte nahe dem Gipfel' leaves it doubtful, whether he means a solitary specimen or a species of pine.

On his way from the north-western part of the terrain in question to the Alberca, he crossed several walls of lava and small plateaux, deeply covered with volcanic sand, upon which 'flourished a mighty vegetation of well-leaved thorny acacias and luxuriant grasses'. The Volcancito del Norte appeared as a conical hill of ashes, surrounded by deep sand. The latest outflow of lava from the main volcano was 'pitch-black, without any vegetation, partly mixed with red earthy and black sandy substances which are more favourable for vegetation'. He remarks upon the great contrast between this dead mass of lava on the one side of the footpath and the beautiful and varied verdure on the other. The sandy fields of the Alberca were under cultivation of water-melons. Here the eastern extent of the lava flow was partly covered with shrubs and grass.

The ascent of the volcano was made from its eastern shoulder.

To reach the crater from the foot of the cone took ¾ of an hour's stiff climbing, in which we were much assisted by the numerous little trees, the dense shrubs and the high grass, which grew upon and between the blocks of lava and the volcanic rubble. This vegetation is composed mainly of the so-called *Tacote*, a tall shrub with large, rough, oak-like leaves; *Palo tepecuaje*, a tree like acacia with long, broad seed-pods; *Copal*, a thorny shrub with small leaves, which is the commonest and characteristic product of this volcanic soil; *Palo jiote*, a tree of stunted growth with red peeling bark on the stem; *Apanicua*, a tree with pretty, yellow, large flowers, but at the time leafless; lastly the narrow-leaved wild *Maguey*. Various birds made themselves heard, and even some roedeer (*Cariacus toltecanus*) and foxes were seen.

On the north-eastern side of the crater, about 20 steps below its rim, I was startled by the condition of the ground, which was warm and soft,...although sparsely covered with grass, probably due to steam escaping through some cleft beneath....That the crater has sunk in, and therefore also widened since the last eruption, is indicated not only by the amphitheatre-like appearance of its terraces but also by the fact that here and there some tree or shrub is growing upon the inner side of the crater walls, which has slid down with the ledge from the outer rim, especially since, but for these trees, the crater is still absolutely bare of vegetation. No doubt in time its inside will be covered with the same luxuriant growth which is now apparent upon many of the streams of lava which surround the volcano.

M. Orozco y Berra. 'Jorullo (Volcan de).' *Diccionario de Historia y Geografia*, t. IV, año de 1854, p. 453.

In this article there were published for the first time the two reports by Sáyago, dated October 8th and November 13th, 1759. They were obtained from a dossier in the Archivo General y Publico de la Nación.

H. de Saussure. 'Notes sur la formation du Volcan de Jorullo, Mexique.' *Bulletin Soc. Vaudoise des Sciences naturelles*, Séance de 22 Juin, 1859.

Unfortunately de Saussure has never published his intended memoir, in which this accomplished observer would most likely have given a description of the state of the vegetation, etc. The following passages are extracted from his preliminary notes:

> Les nappes de lave, ou malpays, ne sont autres que de vastes écoulements de matières incandescentes qui ont tapissé la vallée, en formant des golfes et des promontoires, comme le ferait une masse de plomb fondu qu'on verserait sur une surface rugueuse. Les bords du malpays élevés de 30 à 80 pieds, ne sont pas une tranche de soulèvement, mais seulement le culot terminal des coulées de lave... il n'y a pas trace de soulèvement de couches selon cet axe [the row of volcanoes], ce qui semble prouver que la pression volcanique n'a pas rompu les couches sous-jacentes, mais qu'elle s'est simplement fait jour à travers une faille par laquelle les matières liquides et fluides ont pu s'échapper... ce n'est pas dans le fond de la vallée [at the foot of the big volcano] que s'est ouverte la crevasse, d'où sont sortis les basaltes qui ont tapissé les districtes environnants, mais c'est au contraire sur le versant oriental de la vallée sur les pentes qui s'abaissent vers le fond de cette dernière, et l'axe volcanique est lui-même parallèle à ce fond.

J. de D. Dominguez. *Boletin de Geografia y Estadistica, Mexico*. New series. T. II, 1870, pp. 561–5.

This contains a copy of Sáyago's reports, and Dominguez certifies the correctness of the copy.

J. Leclercq. 'Une visite au Volcan du Jorullo.' *Bulletin Soc. Géog. Paris*, 1886, pp. 386–402.

A rather journalistic account of a visit made in November, 1883. He notes the exuberant vegetation through which his party had to cut their way during the ascent, e.g. the large-leaved capitaneja, and a snake, four feet long with black and white stripes; when they entered upon the volcanic ashes, all the vegetation ceased, but on the top were parrotillas, tepehuajes and guayabas. A thermometer pushed very far into a cleft at the bottom of the crater recorded 152° F.

J. FELIX und H. LENK. *Beiträge zur Geologie und Palaeontologie der Republik Mexico.* 1889–99.

Felix and Lenk visited the Jorullo in April, 1888. On p. 238 is a woodcut (after a photograph) of the view of the Jorullo from the north-west taken across the palm-studded plain. On p. 239 a picture of 'the only still preserved hornito', with plenty of vegetation. Pl. x, fig. 1, shows a view of the northern margin of the crater without any vegetation. Pl. III is a section from Morelia and Patzcuaro to the Jorullo and thence to the Balsas River. Pp. 31–2: 'in the neighbourhood of the clefts inside the crater, whence the steam escapes, is a vegetation of ferns'.

Felix and Lenk give the absolute height of La Playa, at the bottom of the Malpais near its north-western corner, as 714·7 metres against Humboldt's 788·5 metres.

The highest point of the crater was found in the north-east (Pico de Humboldt), 1232 metres against Burkart's 1214 metres: the next highest was found in the north-west (Pico de Riaño of Ordoñez), 1222 metres, the same as Burkart's 1222·4 metres.

ANDRES VILLAFAÑA. 'El Volcan Jorullo.' *Parergones del Instituto Geologico de Mexico,* t. II, num. 3, Mexico, 1907, pp. 73–130 and 8 plates.

Sr Villafaña, mining engineer and geologist of the Geological Institute, made a survey of the Jorullo in 1906. His paper contains a bibliography, a topographical description, geological and petrographical descriptions, with analyses, of the various rocks, and reprints of Alcedo's account, of Sáyago's reports and of Landivar's poem.

The 8 plates give profiles of the road from Patzcuaro and of the volcanoes themselves; a geological and hypsometric map of the Jorullo, a view after a photograph taken at La Playa and photographs of details of the lava.

Ezequiel Ordoñez. 'Le Jorullo.' *Guide des excursions du Xe Congrès géologique international.* Mexico, 1907. 55 pages, 1 map, geological and hypsometric, of Jorullo and 11 plates with photographic views of the volcanoes, the crater, a bomb, and aspects of the lava.

Sr Ordoñez, of the Geological Institute, established himself with Sr Villafaña at the Jorullo for its proper survey, the first which has ever been made excepting the hastily composed, yet wonderfully good map by Schleiden.

B. Hobson. 'An Excursion to the Volcanoes of Nevada de Toluca and Jorullo in Mexico.' *Geological Magazine*, 1907 pp. 5–13, pl. II.

H. M. Cadell. 'Some old Mexican Volcanoes.' *Scottish Geographical Magazine*, vol. XXIII (1907), pp. 281–312, with 12 plates, 2 maps, and other illustrations in the text.

Both these papers contain accounts of an excursion to Jorullo undertaken by members of the International Geological Congress, which met in Mexico in 1906. Cadell's paper includes a photographic reproduction of Humboldt's sketch, map and section of Jorullo, a copy of Ordoñez' geological map, and several photographs and sketches. Hobson's paper also includes a photograph of Jorullo.

www.ingramcontent.com/pod-product-compliance
Ingram Content Group UK Ltd.
Pitfield, Milton Keynes, MK11 3LW, UK
UKHW040656180125
453697UK00010B/219